GENETICALLY MODIFIED CROPS AND AGRICULTURAL DEVELOPMENT

PALGRAVE STUDIES IN AGRICULTURAL ECONOMICS AND FOOD POLICY

Series Editor: Christopher B. Barrett, Cornell University, USA.

Agricultural and food policy lies at the heart of many pressing societal issues today, and economic analysis occupies a privileged place in contemporary policy debates. The global food price crises of 2007–8 and 2010–11 underscored the mounting challenge of meeting rapidly increasing food demand in the face of increasingly scarce land and water resources. The twin scourges of poverty and hunger quickly resurfaced as high-level policy concerns, partly because of food price riots and mounting insurgencies fomented by contestation over rural resources. Meanwhile, agriculture's heavy footprint on natural resources motivates heated environmental debates about climate change, water and land use, biodiversity conservation, and chemical pollution. Agricultural technological change, especially associated with the introduction of genetically modified organisms, also introduces unprecedented questions surrounding intellectual property rights and consumer preferences regarding credence (i.e., unobservable by consumers) characteristics. Similar new consumer concerns have emerged around issues such as local foods, organic agriculture and fair trade, even motivating broader social movements. Public health issues related to obesity, food safety, and zoonotic diseases such as avian or swine flu also have roots deep in agricultural and food policy. And agriculture has become inextricably linked to energy policy through biofuels production. Meanwhile, the agricultural and food economy is changing rapidly throughout the world, marked by continued consolidation at both farm production and retail distribution levels, elongating value chains, expanding international trade, and growing reliance on immigrant labor and information and communications technologies.

In summary, a vast range of topics of widespread popular and scholarly interest revolve around agricultural and food policy and the economics of those issues. This series features leading global experts writing accessible summaries of the best current economics and related research on topics of widespread interest to both scholarly and lay audiences.

The Economics of Biofuel Policies: Impacts on Price Volatility in Grain and Oilseed Markets
   by Harry de Gorter, Dusan Drabik, and David R. Just

Genetically Modifed Crops and Agricultural Development
   by Matin Qaim

GENETICALLY MODIFIED CROPS AND AGRICULTURAL DEVELOPMENT

Matin Qaim

First published 2016 by
PALGRAVE MACMILLAN

The author has asserted their right to be identified as the author of this work in accordance with the Copyright, Designs and Patents Act 1988.

Palgrave Macmillan in the UK is an imprint of Macmillan Publishers Limited, registered in England, company number 785998, of Houndmills, Basingstoke, Hampshire, RG21 6XS.

Palgrave Macmillan in the US is a division of Nature America, Inc., One New York Plaza, Suite 4500, New York, NY 10004-1562.

Palgrave Macmillan is the global academic imprint of the above companies and has companies and representatives throughout the world.

Hardback ISBN: 978–1–137–40571–5
E-PUB ISBN: 978–1–137–40573–9
E-PDF ISBN: 978–1–137–40572–2
DOI: 10.1057/9781137405722

Distribution in the UK, Europe and the rest of the world is by Palgrave Macmillan®, a division of Macmillan Publishers Limited, registered in England, company number 785998, of Houndmills, Basingstoke, Hampshire RG21 6XS.

Library of Congress Cataloging-in-Publication Data

Qaim, Matin, author.
    Genetically modified crops and agricultural development / Matin Qaim.
        pages cm.—(Palgrave studies in agricultural economics and
    food policy)
    Includes bibliographical references and index.
    ISBN 978–1–137–40571–5 (hardback : alk. paper)
    1. Transgenic plants. I. Title. II. Series: Palgrave studies in agricultural
    economics and food policy.

SB123.57.Q35 2015
631.5′233—dc23   2015017986

A catalogue record for the book is available from the British Library.

CONTENTS

FIGURES AND TABLES

Figures

FOREWORD

Population and income growth, combined with continued urbanization, will result in broad dietary change and a doubling of food demand in today's developing countries by 2050. Meanwhile, climate change will pose new biotic and abiotic challenges to food production. So will rising concerns in the high-income countries about the environmental footprint of modern agriculture. Consumers, meanwhile, increasingly want and are willing to pay for specific product attributes, both substantive ones, like enhanced mineral or vitamin content, as well as aesthetic ones like uniform color and shape. How will the world meet these supply and demand side challenges in the decades ahead?

To many scientists and policymakers, genetically modified (GM) crops and livestock offer an important part of the answer. But many consumers and environmental groups oppose these new technologies. Indeed, the battles over GM foods have arguably been among the most controversial topics in global agriculture over the past 20 years. The considerable potential of modern methods of genetic modification to accelerate the adaptation of animals and plants to evolving environmental conditions and consumer tastes offers historically unprecedented opportunities to increase agricultural productivity, improve yield stability, and reduce the use of agrochemical inputs. But the intense popular reaction against GM crops in some countries, especially in Europe, underscores that science does not always have the final word in policy debates.

In this book, Matin Qaim, one of the world's foremost experts on the economics of genetically modified crops, meticulously reviews the evidence on GM crops within the context of developing countries, where the battle lines are perhaps most stark and the stakes highest. He carefully walks us through the now-considerable evidence that GM crops are not intrinsically more risky than conventionally bred crops or other agricultural technologies. He documents the dramatic diffusion of GM crops since the mid-1990s, when they first became widespread, mainly in North America. By 2014, 182 million hectares worldwide were sown with GM seeds, more than half of this area in developing countries. While the popular debates about GM crops have raged, the developing world has quietly become the global leader in GM agricultural production. Qaim summarizes the growing body of scientific evidence that clearly indicates that GM crops have overwhelmingly benefitted farmers, consumers, and the environment, in spite of many (scientifically unsupportable) popular claims to the contrary.

Dr. Qaim has been working on these topics since the 1990s, even before the GM debates began to regularly take over the front pages of newspapers. With 20 years' accumulated expertise built from carefully studying agricultural biotechnology and GM crops across a range of sub-sectors and countries, he deploys his formidable technical skills and depth of knowledge to make clear for readers the key issues in these debates. In an extremely technical area where noise too often dwarfs signal, Qaim provides a concise and accessible overview of the broad literature about economic and social dimensions of GM crops. He analyzes whether GM crops can contribute to sustainable agricultural development and what types of policies are required to optimize the benefits and to avoid undesirable outcomes. He provides many interesting examples and puts GM crops into the historical context of other breeding methods and earlier technological breakthroughs in agriculture. He explains not only the economic research on the impacts of GM crops but even the basic tools of molecular breeding in a way that non-experts can easily grasp.

Of particular interest, he draws on his own research group's and others' extensive, rigorous research to demonstrate that poor farmers and consumers typically benefit substantially from GM crops. The GM crops commercialized so far already contribute to productivity and income gains in the small farm sector, helping to reduce poverty and improve food security. The potential welfare effects of future GM technology applications are much larger still. Nonetheless, most of the poorest countries in Africa and Asia have not yet approved GM crops, especially not food crops, as cotton has been the dominant GM crop cultivated thus far in the developing world. Qaim explains how the release and diffusion of promising technologies is too often impeded by excessive regulatory hurdles and negative propaganda by anti-biotech activists. He shows convincingly that public attitudes and policies related to GM crops in Europe and other developed countries also have a profound influence on what happens in the developing world.

Qaim takes us through the complex web of policy and regulatory issues related to biosafety and food safety, intellectual property rights, industry structure, international trade, and food labeling, among other topics. With substantial insider knowledge he discusses many of the public misconceptions and explains why they persist in spite of mounting evidence to the contrary. The underlying political economy is a fascinating story, as several groups directly benefit from the global protest movement against GM crops. Dr. Qaim concludes that better science communication and more integrity in public and policy debates are required if the developing world is to realize the considerable potential of GM crops to advance food security and broader socioeconomic development objectives.

For those unfamiliar with the academic research to date on the broader societal effects of GM crops, I can think of no better scholar to introduce this hot-button topic than Matin Qaim. In these pages he offers an extremely clear, careful treatment of a complex issue. I learned a great deal from reading it. I highly recommend this book as an essential reference about one of the most important topics in agricultural economics and food policy in the early twenty-first century.

CHRISTOPHER B. BARRETT
Cornell University

PREFACE

I started working on economic and social aspects of genetically modified (GM) crops in 1996 as part of my doctoral thesis research. Working on this topic was not my own idea. I had studied agricultural sciences and agricultural economics and was eager to do research related to hunger and poverty in developing countries. When my doctoral thesis advisor, Joachim von Braun, suggested working on agricultural biotechnology and genetically modified organisms (GMOs) I was really hesitant in the beginning. I did not know much about GMOs at that time, but I was skeptical. There were a couple of student groups I sympathized with that were strongly opposed to GMOs. I had heard about environmental, health, and social risks and the fact that private companies, including a few multinational corporations, were dominating the development of GM crops. I did not see much potential of this technology to contribute to poverty reduction in developing countries. I was also somewhat afraid of my friends frowning upon me when I would tell them that I worked on GMOs. My doctoral thesis advisor agreed that I could also work on other topics, but after some more discussion he convinced me that the biotech direction is really interesting, as almost nothing was known about the wider implications for the poor. So I decided to concentrate on this direction for a couple of years.

As a newcomer to the biotech topic I read a lot, both scientific and less-scientific papers and books. I also attended a number of scientific meetings, policy workshops, and public hearings where the pros and cons of GMOs were discussed, often emotionally. Sometimes there were developing country farmer representatives flown in for these meetings upon invitation from German NGOs. Most of these farmer representatives were really eloquent. They all stated how much they hated GMOs because this technology would destroy biodiversity and traditional knowledge systems in developing countries. I was impressed when I heard the first such speech by a so-called farmer representative. Additional speeches rather made me suspicious; all of them were very similar, regardless of where the speakers came from and the fact that GMOs were not used in any of their home countries at that time.

When I started my own field research and data collection in developing countries, I got a very different picture. Unsurprisingly, farmers that I interviewed typically knew nothing about the science of GM crops or their effects on biodiversity, but all of them were eager to try new seed technologies that could help address some of their pressing agronomic problems, as long as these new seeds would be available at affordable prices. I also met numerous biotech scientists, plant breeders, agronomists, ecologists, and extension officers in various countries and learned a lot about their work and perspectives. More and more I realized how powerful GM technology could be and how much it could contribute to rural development, when the research priorities are set accordingly. I saw an important role for the public sector, because multinationals alone would not address the technological needs of smallholder farmers in developing countries. I also recognized that improving national research capacities, rural infrastructure, and smallholders' access to markets are important preconditions for equitable technological development. Even scale-neutral technologies can aggravate inequality when access to these technologies is uneven.

Initial studies that I carried out on GM crops were *ex ante* impact assessments. Based on research and experimental results, expert statements, and detailed data about the given farming conditions in a particular country, I simulated how GM technology adoption and impacts might develop in the future under different policy assumptions. Later, when GM crops were increasingly commercialized and adopted in developing countries, I focused on *ex post* studies, collecting and analyzing data from randomly sampled farmers that I and my students surveyed, sometimes repeatedly over various years to also understand the underlying dynamics. Over the last 20 years, together with my research group we collected comprehensive survey data on GM crop aspects in various developing countries, including Argentina, Brazil, India, Kenya, Mexico, Pakistan, and the Philippines. I also had the chance to talk to farmers, researchers, and policymakers about issues of agricultural biotechnology in several additional developing countries, including China, Ethiopia, Indonesia, South Africa, Tanzania, Thailand, and Vietnam.

When I started working on GMOs in the mid-1990s, I did not expect that this topic would remain one of my major research areas for the next 20 years, and possibly beyond. I am not a natural scientist with a biotech lab and unique research experience on particular molecular techniques. For agricultural economists, it is quite common to work on certain topics for a few years and then switch to other topics where new interesting issues emerge. Over the years, I have started working on various other topics related to agriculture, nutrition, and food systems in developing countries, but I decided to also continue my work on the economics of biotech. Having an applied and policy-oriented focus, I was never satisfied by publishing academic papers alone. I also wanted to see that the knowledge generated through the research of my group and many other colleagues would enter the public debate and eventually contribute to more informed and science-based policymaking. Unfortunately, this has not yet happened. I am deeply troubled by the fact that the public GMO debate in Europe is completely detached from the scientific evidence accumulated over the last 30 years. I cannot deny that this is frustrating at times, but I also take this as a sign that the work is not yet done. This is also why I agreed to write this book when I was approached by Chris Barrett and Palgrave Macmillan.

When I prepared my first lecture on issues of agricultural biotechnology almost 20 years ago, I had designed a slide (an overhead transparency at that time) listing the most common arguments for and against GM crops that were regularly used in the public debate at that time. This is not remarkable. More remarkable is that I still use exactly the same slide to motivate my lectures today, and this slide still accurately summarizes the current state of the public debate. The arguments have not changed at all. The only difference is that today more people in the lecture audience believe that the listed concerns have become true, while the listed arguments about potential benefits have remained empty promises. These public perceptions reflect the opposite of what happened in reality. There is now strong evidence that GM crops are beneficial for farmers, consumers, and the environment, and that they are as safe as their conventionally bred counterparts. In this book, I give an overview of what we know about the impacts of GM crops and their wider repercussions. I also discuss where I see shortcomings and need for public action. Finally, I try to explain why scientific evidence about GMOs had so little influence on public perceptions in Europe and elsewhere. I hope this book will not only be read by the same old participants in the biotech debate with their entrenched views but can also reach out to a broader open-minded readership that is willing to take a fresh perspective.

I do not make an attempt to hide that my views have changed and that I now see great potential in GM crops to contribute to agricultural development. My assessment is not based on any preconceived opinion, but on 20 years of studying and carrying out own research on this topic in various parts of the world. I do not develop GM crops myself and therefore have no vested interest in finding positive, negative, or no impacts of this technology at all. My motivation is entirely driven by the question whether, and, if so, how GM crops can contribute to sustainably increasing agricultural productivity, reducing poverty, and improving food security. I am convinced that the world is better off with GM crops than without, and that future challenges of agricultural development can only be properly addressed if we harness all promising areas of science responsibly. Once the ideological rejection of GMOs is overcome, which I am still optimistic will happen at some point, the debate and protest energy should concentrate much more constructively on what needs to be done to optimize the social benefits. Like for any transformative technology, institutional and policy adjustments are necessary to fully reap the potentials and avoid undesirable consequences.

Researchers who find positive effects of GM crops are sometimes accused of being influenced by corporate interests. I would like to stress that my research on GM crops was never influenced by corporate interests and never funded by industry money. Most of my research projects over the last 20 years were funded through competitive research grants obtained from the German Research Foundation (DFG). The rest was funded by several other public sector organizations and philanthropic foundations, including the German Federal Ministry of Economic Cooperation and Development (BMZ), the EU Commission, USAID, the Rockefeller Foundation, and the Eiselen Foundation (now Foundation fiat panis). I gratefully acknowledge this financial support for my research. I would also like to thank the University of Goettingen, where I have been working for several years now and always get the necessary support and freedom for my research. Before moving to Goettingen, I carried out GM crop related research at the University of Bonn, the University of California at Berkeley, and the University of Hohenheim in Stuttgart. I also thank these organizations for providing support and stimulating academic environments.

Over the last 20 years, I have learned a lot from many people who influenced my thinking about GM crops and agricultural development. I benefited tremendously from cooperating with extraordinary scholars and practitioners in this field. In particular, I would like to mention Arnab Basu, Peter Beyer, Howarth Bouis, Alain de Janvry, Clive James, Anatole Krattiger, Tom Lumpkin, J. V. Meenakshi, Michael Njuguna, Ingo Potrykus, Carl Pray, N. Chandrasekhara Rao, Joachim von Braun, Florence Wambugu, Usha Barwale Zehr, and David Zilberman. I would also like to thank the doctoral and postdoctoral researchers who worked with me on issues of agricultural biotechnology at the Universities of Bonn, Hohenheim, and Goettingen. In particular, these were Abedullah, Carolina González, Jonas Kathage, Wilhelm Klümper, Shahzad Kouser, Vijesh Krishna, Ira Matuschke, Prakash Sadashivappa, Alexander Stein,

Arjunan Subramanian, Prakashan Chellattan Veettil, and Roukayatou Zimmermann. These scholars spent a few years in my group and took up exciting positions elsewhere in the world after finishing their doctoral degrees or postdoc sojourns. All of them had brilliant ideas and contributed to the success and visibility of my group. Useful research assistance for this book was provided by Markus von Kameke.

Finally, I would like to thank my family for always supporting me in my work on a controversial topic. My wonderful wife Christina is always a great source of inspiration and personal advice. And my two marvelous daughters showed interest in the topic, but were also happy when I told them that I completed the manuscript. I dedicate this work to my three beloved ladies, Christina, Charlotte, and Lina.

CHAPTER 1

INTRODUCTION

What are the goals and priorities of agricultural development? Answers to this question can be diverse. Depending on who is being asked, the list of priorities may include food security, poverty reduction, supply of biofuels, soil conservation, biodiversity preservation, climate protection, animal welfare, attractive rural landscapes for recreation, and many other things. People in Western Europe will likely answer differently from people in South Asia or sub-Saharan Africa because of different living standards, cultural backgrounds, and attitudes. Also within regions, priorities may differ between rich and poor, urban and rural, young and old, men and women, and so on. Moreover, responses to the question about goals and priorities today would probably be quite different from responses 20 or 50 years ago. However, in spite of the many nuances and changes in priorities and preferences over time, there are a few overarching goals of agricultural development that persist and that constitute the foundation for this book. I focus on three goals in particular and shall analyze how far genetically modified (GM) crops can contribute to achieving these goals.

The first goal of agricultural development is to produce sufficient food and other agricultural commodities to satisfy the needs and preferences of the growing human population. This does not mean that growth in agricultural supply has to match growth in demand everywhere because international trade can help to balance disequilibria between surplus and deficit regions. National food self-sufficiency is usually not an efficient objective because population growth and endowments of land, water, and other natural resources required for agricultural production differ geographically. Globally, however, sufficient production is an important precondition for food security—defined as every person having access to sufficient and nutritious food to maintain a healthy and active life. If the growth in agricultural demand is higher than the growth in supply at the global level, prices will rise, making food less accessible for the poor.

The second goal is to improve the livelihoods of the people directly involved in the agricultural sector, including farmers and farm workers. With overall economic development, the proportion of people active in agriculture shrinks, as the industrial and services sectors gain in importance. This normal structural change should not be obstructed. However, in many developing countries agriculture is still the most important source of employment, especially for the poor. Around three-quarters of all the poor and undernourished people worldwide live in rural areas and derive a large share of their income from agriculture (World Bank, 2013). Many of the poor are small-scale farmers. Hence, agricultural growth in the small farm sector is an important avenue for poverty reduction and improved nutrition.

The third goal is related to sustainability. Sustainability requires natural resources and the environment to be preserved, so that humanity will be able to achieve the first two goals also in the long run. This underlines the close interconnection between the three overarching goals of agricultural development.

The last few decades have seen remarkable progress toward the first goal. Growth in agricultural production outpaced population growth. Historically, increases in agricultural production were primarily achieved by using additional land. However, over time land became scarcer so the focus shifted toward increasing yields per unit area. Advances in agricultural research and development (R&D)—especially in breeding, plant nutrition, pest control, and engineering—have led to large yield increases in many parts of the world over the last 50 to 60 years. Since the 1960s, the total land used to cultivate crops has hardly increased, while global food production has more than tripled. The observed production increase was primarily due to farmers switching from traditional landraces to new high-yielding crop varieties and using more fertilizers, chemical pesticides, and methods of irrigation.

Progress toward the second goal of agricultural development was also remarkable during the last few decades. While hunger and poverty are still widespread in rural areas of Asia and Africa, the proportion of poor people has declined considerably. In 1950, more than half of the world population lived in extreme poverty, compared to around 15 percent in 2010 (United Nations, 2014). Poverty reduction is the result of many factors, including improvements in education, infrastructure, and social services. Agricultural R&D and the implementation of new technologies in the small farm sector have also played a significant role (Eicher and Staatz, 1998; Thirtle et al., 2003; Fan et al., 2005; World Bank, 2007).

Progress toward the third goal of agricultural development—sustainability—was much more mixed during the last 50 to 60 years. On the one hand, the yield increases on the cultivated land have helped to reduce cropland expansion to forests and other pristine areas (Evenson and Gollin, 2003; Villoria et al., 2014), thus preserving natural biodiversity and reducing greenhouse gas emissions from additional land use change. On the other hand, the intensification of agricultural production and a sharp increase in the use of agrochemicals have brought about other environmental problems, such as soil degradation, emission of nitrous oxides, contamination of water with toxic residues, and loss of biodiversity in farming environments. The replacement of a large number of landraces with a smaller number of high-yielding crop varieties may also have contributed to agrobiodiversity erosion (Tripp, 1996).

Addressing these environmental problems remains a challenge for agricultural development. Many argue that the use of external inputs has to be drastically reduced or avoided completely to ensure environmentally friendly production. In the public discourse, some groups equate sustainable agriculture with organic production methods, which—they argue—needs to be scaled up from its current niche position. Certified organic agriculture, currently covering less than 1 percent of the world agricultural land, builds on ecological principles and rules out the use of mineral fertilizer and chemical pesticides (FiBL and IFOAM, 2014). But is a reduction of agrochemicals always good from a sustainability perspective? Regional differentiation is required. In Western Europe and the United States, the use of chemical fertilizers and pesticides is relatively high, but has declined since the 1990s. Today, according to data from the Food and Agriculture Organization (FAO), farmers in the United States apply 130 kg of mineral fertilizer per hectare of cropland on average. Farmers in Germany use around 200 kg per hectare. Further reductions from such levels may be desirable to contribute to more environmentally friendly production systems. In a few other countries, much higher amounts of agrochemicals are being used. In China, for instance, farmers apply around 650 kg of mineral fertilizer per hectare, causing much more significant environmental problems that need to be addressed. On the other hand, in many countries of sub-Saharan Africa less than 10 kg of fertilizer is used on average. Soils in Africa are often severely nutrient-depleted. In such situations, a further reduction in fertilizer use would not contribute to more sustainable production. On the contrary, increasing the fertilizer use could not only increase yields but also contribute to environmental benefits, as the pressure of agricultural expansion to ecologically fragile areas would be reduced. These examples demonstrate that there are no one-size-fits-all solutions for making agricultural production systems more sustainable.

Beyond reducing the environmental footprint of production, other challenges for agricultural development remain. The progress made over the last decades in terms of poverty and hunger reduction should not lead to complacency, as the agenda is not yet finished. The FAO estimates that close to eight hundred million people are still undernourished, meaning that their access to and intake of calories is insufficient (FAO, 2015a). But healthy nutrition is not about calories alone. Around two billion people worldwide suffer from deficiencies in one or multiple micronutrients—such as iron, iodine, zinc, or vitamins—with serious negative health effects (IFPRI, 2014). And the demand for food and feed increases due to population and income growth. In addition, demand is driven by the increasing use of agricultural products for bioenergy and other industrial purposes. Long-term projections are always associated with some uncertainty because changing preferences and the role of policy cannot be perfectly predicted. An international team of researchers has reckoned that global agricultural production may have to double between 2010 and 2050 to keep pace with the rising demand for food, feed, fiber, and biofuel (Godfray et al., 2010). Projections by the FAO and other organizations are in a similar range (Giddings et al., 2013). Reducing food losses and waste along the value chain is also an important objective that needs to be pursued. But even if losses can be reduced, a production challenge will remain; it is not an "either-or" question. Global agricultural production will have to be increased considerably over the next couple of decades to ensure sufficient food availability in the future (Foresight, 2011; Oxfam, 2011; Rosegrant et al., 2014; Hertel, 2015).

How can agricultural production be increased sustainably when natural resources are becoming increasingly scarce? Expanding the agricultural land may be possible in some regions, but additional land use change is associated with environmental costs in terms of greenhouse gas emissions and potential biodiversity loss. Hence, as was true already in recent decades, the main part of the required production increase will have to come from higher yields. Using more water, mineral fertilizer, and chemical pesticides may still contribute to higher yields in some regions, especially in Africa, but cannot be the paradigm elsewhere because of the associated environmental problems. Water is also scarce and already overused in many parts of the world. The production of nitrogen fertilizer is very energy-intensive. An additional complexity is climate change, to which agriculture contributes, but which is also affecting agricultural production potentials. While agriculture in a few world regions may benefit from rising temperatures, significant negative effects are predicted in many tropical and subtropical regions (IFPRI, 2010; Foresight, 2011). Added heat and drought stress, as well as more frequent weather extremes, could reduce crop yields by more than 20 percent in South Asia and sub-Saharan Africa, if suitable adaptation strategies cannot be found and implemented.

The main route of increasing agricultural production sustainably is not through using more natural resources but through developing and deploying improved technologies that help to reduce the environmental footprint per unit of production. In the past, new technology often involved high-yielding crop varieties coupled with more chemical inputs and irrigation. In the future, approaches have to be different. Yield increases will remain central, but ways have to be found to loosen the correlation between yield and external input use, and to make production systems more resilient to environmental stresses. Different expressions have recently been established to describe such kinds of agricultural innovation. The Royal Society (2009) has coined the term "sustainable intensification." "Sustainable agriculture" and "natural resource management" technologies are somewhat older terms but with similar concepts (Lee, 2005). More recently, the term "climate-smart agriculture" has become popular (FAO, 2013). Different groups of people use these terms sometimes with different priorities in mind, but this can be misleading because there is a close overlap in the definitions (Godfray, 2015). Sustainable production systems require locally adapted combinations of improved seeds, improved agronomy, engineering, and information technology. In this book, the focus is on plant breeding, and GM crops in particular, but it should be stressed that GM crops cannot substitute for the other types of innovations and practices required to make production systems sustainable.

Plant Breeding and GM Crops

Plant breeding significantly contributed to yield increases in the last 100 years, and its role has increased over time. Based on data from various world regions, Evenson and Gollin (2003) estimated that between 1960 and 1980 around 20 percent of the yield gains in major cereals were directly attributable to improved seeds. The rest was primarily due to increases in the use of irrigation, chemical inputs, and machinery. Between 1980 and 2000, the contribution of improved seeds had increased to 50 percent because of diminishing returns to other inputs. Conventional breeding is also subject to diminishing returns, as cross-breeding relies on the existing genetic variability within a particular crop species. For long, breeders have tried to increase this genetic variability through crosses with wild relatives, hybridization, induced mutations, and other approaches. Modern biotechnology is offering new tools to improve the breeding efficiency, without necessarily changing the breeding objectives. But the options to develop crop plants with desirable traits have certainly increased. A better understanding of the genetic makeup of plants has enabled the analysis of gene locations and their functions. Individual genes can also be isolated from one organism and transferred to the cells of another organism. This gene transfer is possible between organisms of the same species or also across species boundaries. Thus, the genetic variability available to develop desirable traits in plants has vastly increased. Using cell and tissue culture techniques, whole plants can be regenerated from the cells into which the desired genes have been introduced. With these new biotech tools, breeding has become much more targeted and precise.

A GM crop is a plant used for agricultural purposes into which one or several genes coding for desirable traits have been inserted through genetic engineering. The basic techniques of plant genetic engineering were developed in the early 1980s, with the first GM crops becoming commercially available in the mid-1990s. Since then, GM crop adoption has increased rapidly. In 2014, GM crops were already grown on 182 million hectares, equivalent to 13 percent of the global arable land (James, 2014). With this wide coverage within a relatively short period of time, GM crops are among the fastest-adopted agricultural technologies in human history. However, adoption patterns are geographically very uneven. While farmers in North and South America and a few countries in Asia have rapidly embraced GM crop technologies, adoption in Europe and Africa is still very low, due to various reasons.

As mentioned, the crop traits targeted through genetic engineering are not completely different from those pursued by conventional breeding. However, since genetic engineering allows the direct transfer of genes across species boundaries, some traits that were previously difficult or impossible to breed, can now be developed with relative ease. Three categories of GM traits can be distinguished. The so-called first-generation GM crops involve improvements in agronomic traits, such as better resistance to pests and diseases. Second-generation GM crops involve enhanced quality traits, such as higher nutrient contents of food products, while third-generation crops are plants designed to produce special substances for pharmaceutical or industrial purposes (Qaim, 2009; Kempken and Jung, 2010).

The potentials of GM crops to contribute to agricultural development are manifold. Plants that are more resistant to pests and diseases, and more tolerant to abiotic stress factors such as drought and heat, could enable higher harvests and more yield stability, while reducing the reliance on chemical pesticides and irrigation water. Plants that use soil nutrients more efficiently could contribute to higher yields with lower mineral fertilizer. And plants that contain higher amounts of micronutrients in their edible parts could help to reduce nutritional deficiencies and thus improve human health. While all of these traits are being developed by plant researchers, and many have already been tested in the field, only a few GM traits in a small number of crop species have so far been approved and released for practical use by farmers. Most of the commercial GM crop applications so far involve herbicide tolerance and insect resistance in soybean, maize, cotton, canola, and a few other crops (James, 2014). The evidence so far suggests that these early applications of GM crops have contributed to significant productivity gains and environmental benefits in agricultural production (Qaim, 2009; Carpenter, 2010; Finger et al., 2011; Areal et al., 2013; Klümper and Qaim, 2014).

Limited Public Acceptance

In spite of the potentials of GM crops to contribute to agricultural development, their introduction has aroused significant opposition (Gilbert, 2013). The intentional transfer of genes across species boundaries is considered highly unnatural by many, causing ethical concerns. This is a difficult debate because it is hard to define where "natural" ends and where "unnatural" begins. In nature, the exchange of genetic information primarily happens through cross-fertilization of individual organisms within one species (sexual reproduction), although spontaneous horizontal gene transfer across species boundaries also occurs. It is not uncommon to find plants containing genetic sequences from microorganisms that were transferred naturally through plant–microbe interactions (Kyndt et al., 2015). Even humans carry foreign genes from algae, fungi, bacteria, and other species that immigrated to the human genome at some point in the evolutionary history and were passed on to the offspring since then. Recent research has shown that humans have picked up at least 145 genes from other species during the course of evolution (Crisp et al., 2015).

It is clear that the GM crops that have been developed and commercialized would not have emerged naturally without human intervention, but the same holds true for all conventionally bred crops as well. The domesticated crops that are widely used in agricultural production today are very different from their natural ancestors because of millennia of human selection and breeding. In this sense, all technologies that humans have developed are unnatural. Of course, genetically modified organisms

(GMOs) are different from technologies in the automotive or computer industries, as living organisms are directly involved. Research on living organisms is usually associated with different types of ethical concerns. But living organisms are also involved in the genetic engineering of bacteria and other microorganisms that are widely used in food processing and for the production of medical drugs. Many drugs that are widely used today were developed with the help of genetic engineering, without much public debate about ethical concerns.

Beyond ethical aspects, there are widespread concerns about health and environmental risks of GM crops. Since the complex functions and interactions of genes are not yet fully understood, it is feared that introducing new genes might possibly cause the emergence of substances that are toxic to humans or other nontarget organisms. There are also worries that the introduced genes might outcross to wild relatives of the domesticated crops, possibly causing biodiversity erosion or other ecosystem disruptions. Concerns about environmental and health risks have led to complex biosafety and food safety regulations. While the concrete regulatory approaches and responsible authorities differ between countries, international agreements require that GMOs cannot be released without comprehensive risk analysis, testing, and approval by the regulatory authorities. The regulatory hurdles are much higher for GM crops than for any other agricultural technology. This is in spite of the fact that there is no evidence that GM crops have greater adverse impact on health and the environment than other crops developed by alternative technologies used in plant breeding (EASAC, 2013; House of Commons, 2015).

There are also public concerns about possible adverse social implications of GMOs, especially when it comes to the use of this technology in developing countries (Glover, 2010). For instance, some believe that GM technology could undermine traditional knowledge systems in local communities. Given the increasing privatization of crop improvement research and the proliferation of intellectual property rights (IPRs) there are also concerns about the potential monopolization of seed markets and exploitation of farmers (Shiva et al., 2011). Almost all GM crops commercialized so far were developed by private companies, primarily large multinationals. Even when exploitation should not be an issue, it is questionable whether multinationals would focus on the needs of smallholder farmers in terms of R&D priorities and GM seed supply. If only large farms were to use GM crops, existing inequality would rise and small farms would be further marginalized.

The multinationals that are developing and commercializing GM crops—such as Monsanto, Pioneer/DuPont, Syngenta, Bayer

CropScience, Dow AgroSciences, or BASF—all happen to be companies with a business background in the agrochemical pesticide industry. This fact does not necessarily help to create trust in GM crops and their environmental friendliness. While chemical pesticides have an important role to play for agricultural development, their public image is rather bad. In her famous book *Silent Spring* that was published in the early 1960s, Rachel Carson reported disastrous environmental and health effects of the indiscriminate use of synthetic pesticides and accused the chemical industry of spreading disinformation. This book became very influential in the global environmental movement, so that pesticides and the companies producing them are seen by many as a major environmental evil. Claims that GM crops might reduce chemical pesticide use and contribute to sustainable development do not seem to be very convincing when these crops are primarily developed by the same companies. Such deep-rooted distrust complicates the discussion because any argument of possible benefits of GM crops is often dismissed as industry propaganda. Even when public sector scientists talk about potentials of GM crop technology, there is immediate suspicion that these scientists must have been influenced by industry money. Instigated by large international non-governmental organizations (NGOs) with an anti-biotech agenda—such as Greenpeace and Friends of the Earth—it has become a public norm that GM crops are undesirable.

Especially in Europe, public perceptions of GM crops are those of a technology that has no obvious benefits, but is risky and unpredictable in terms of its consequences, brings about patenting of life, is dominated by multinational companies, and fosters monopolies and monocultures. According to recent polls, the large majority of the citizens in Western Europe reject GM crops (European Commission, 2010a). In Germany and Austria, over 90 percent of the people state that they could not imagine consuming GM foods. Not all Europeans know that they already consume foods derived with the help of GMOs on a regular basis. While mandatory labeling of GM foods exists in the European Union (EU), foods derived from animals fed with GM crops (e.g., meat, milk, eggs) do not fall under the labeling requirement. The EU imports large quantities of soybean meal used as feed from North and South America, where GM soybeans are widely grown.

Negative attitudes toward GMOs in Europe have to be seen in a wider context. They are part of a broader movement against modern agriculture that—in the views of many—focuses on productivity alone without considering negative environmental and social externalities. The International Green Week, a large agriculture and food exhibition taking place every year in Berlin, is regularly accompanied by NGO-organized street protests where tens of thousands of people demonstrate against "industrial agriculture" under the motto "we are fed up." Wealthy urban consumers are increasingly detached from the realities of agricultural production. Many have a romanticized notion of how agriculture should look like, a notion that is more similar to farming 100 years ago than to modern agriculture today. The need for production and productivity growth is not always recognized by Western urbanites with full stomachs. Hence, any productivity-increasing agricultural technology has a hard time in getting accepted. If on top there are perceived environmental and health risks, serious opposition is almost inevitable.

In the United States, the acceptance of GM crops is generally higher, but recent debates about the mandatory labeling of GM foods in several American states show that public suspicion also exists (The Economist, 2014). European attitudes have spilled over to several developing countries as well (Paarlberg, 2008; The Economist, 2013). Specific fears about risks, intermingled with broader concerns about corporate control of food, have contributed to a global protest movement against GMOs. Almost everywhere, policymakers have become very cautious to approve new GM crop applications. Some countries have essentially banned any new releases of GMOs. Rates of innovation in plant biotechnology nowadays depend much more on public acceptance and regulation than on technological needs and possibilities.

Objectives of This Book

Notwithstanding limited public acceptance, GM crops have been used in many countries for almost 20 years, so that impacts can already be observed. The evidence available suggests that the public concerns about environmental and health risks of GM crops are overrated, while the benefits are underrated. A recent meta-analysis of studies that looked at agronomic and economic impacts of GM crops worldwide showed that this technology has contributed to significant reductions in the use of chemical insecticides and increases in agricultural productivity and incomes, especially for farmers in developing countries (Klümper and Qaim, 2014). There is also evidence of environmental, health, and nutritional benefits (Hossain et al., 2004; Kouser and Qaim, 2013; Qaim and Kouser, 2013; Huang et al., 2015). This does not mean that GM crops have positive effects only, but the negative effects observed in some situations are related to inappropriate use rather than being inherent to GM technology. Unfortunately, the available evidence has hardly entered the public debate. The prejudices and arguments used against GM technology are still the same as 20 years ago.

This book makes an attempt to analyze the potentials and limitations of GM crops from a sustainable development perspective, including economic, social, and environmental aspects. The political economy of the GM crop debate will also be discussed. I review the empirical evidence about impacts of already commercialized GM crops and expected effects of GM technologies that are still in the R&D pipeline. I also discuss risks, regulatory issues, and policy aspects. Finally, I delve into the controversies in the public debate, trying to better understand the concerns and why scientific evidence has not been more successful in moving the debate forward. The objective of this book is to contribute to a more rational discourse about GM crops by providing science-based information on various aspects of public concern. Public opinions are not shaped by scientific evidence alone. But scientific evidence still has an important role to play for dispelling widespread misconceptions.

Overview

The book is structured as follows. Chapter 2 analyzes the role of technology, in general, and plant breeding, in particular, in the history of agricultural development. This historical perspective is important to better understand similarities and differences between GM crops and other breeding technologies. While farmers have selected and exchanged the most promising seeds for replanting since the beginnings of agriculture some 12,000 years ago, modern plant breeding only became possible after the discovery of the basic rules of genetic heredity in the nineteenth century. Systematic plant breeding contributed to unprecedented yield increases in the twentieth century. From today's perspective, the breeding approaches used then are referred to as conventional breeding. In reality, various breeding techniques were used, further developed, and often combined. Hence, the term conventional breeding is very broad and not really useful to describe one particular technique. In addition to a review of some technical aspects of breeding, developments in the seed sector are summarized in chapter 2, including a discussion of the changing roles of public and private sector organizations.

Genetic engineering is a set of additional techniques in the breeding toolbox that can help to further increase the efficiency of developing desirable crop traits. Chapter 3 provides some simple technical background of how GM crops differ from conventionally bred crops. The Chapter also gives an overview of breeding objectives that are currently pursued with GM techniques and their potentials to address agronomic and nutrition constraints. This discussion of potentials is followed by a review of possible environmental, health, and social risks of GM crops.

Chapter 4 provides an overview of the adoption of commercialized GM crops in different regions of the world. It also reviews the literature about GM crop impacts, differentiating between herbicide-tolerant and insect-resistant crops. A growing body of literature has looked at effects of GM crop adoption on crop yields, chemical pesticide use, costs of production, and farmer profits, using different types of data and methodologies. Some studies have used cross-section data, comparing the performance of GM crop adopting and non-adopting farms at one point in time. Other studies have used repeated surveys and panel data to analyze effects over time. Especially in developing countries, research has also examined impacts of GM crop adoption on smallholder farmers' income, occupational health, and poverty. Much of this research has concentrated on insect-resistant cotton, which is the most widely adopted GM crop technology in the small farm sector up till now. Chapter 4 summarizes these results from various countries. In addition, it provides a case study of GM cotton adoption in India. This case study is of particular interest because the debate about impacts of GM cotton on small farms in India has been particularly controversial in recent years, with reported effects ranging from large benefits to disastrous failures.

An overview of the R&D pipeline is provided in chapter 5. Many GM crops and traits not yet used by farmers were already developed and tested in the field, so that they may be commercialized within the next few years. This includes crops with virus and fungal resistance, tolerance to abiotic stress factors—such as drought and soil salinity—as well as technologies to improve nutrient use efficiency, among others. Furthermore, aspects related to GM food crops with higher contents of micronutrients important for human nutrition are discussed. One widely known example of such "biofortified" crops is Golden Rice with high contents of provitamin A in the grain to address problems of vitamin A deficiency. While Golden Rice has been debated widely by GM crop advocates and opponents, this technology has not yet been released for practical use by farmers and consumers. Potential effects of such future GM crop applications are reviewed from an *ex ante* perspective.

In chapter 6, regulatory issues of GM crops are reviewed, including biosafety and food safety regulations, food labeling, coexistence rules, and IPRs. Effects of GM crop regulation on industry structure and innovation rates are also discussed. These are broad and complex topics, all of which would deserve a comprehensive treatment. Discussing all details is beyond the scope of this book. But since trends in GM crop development, commercialization, and impacts cannot be fully understood without some insights into regulatory issues, a summary discussion is important. For further details, the reader is referred to useful other literature sources.

Chapter 7 is devoted specifically to the complex public debate around GM crops, including a discussion of the powerful roles of NGOs in shaping public opinions in Europe and elsewhere. Anti-GMO campaigners have created narratives of fear. These narratives are inconsistent with the empirical evidence of GM crop impacts but are nevertheless perpetuated with the help of the mass media and other stakeholders that benefit from the biotech opposition. Several of these popular narratives are explored and refuted in chapter 7. I also try to explain why entrenched views persist and public perceptions are hardly influenced by the growing empirical evidence about the benefits of GM crops and their safety. The protest movement is seriously stifling further GM crop developments through various channels. Nowadays, new technologies cannot be established more widely when a large majority rejects them. This is an important democratic principle. A problem occurs when public opinions are based on prejudices and biased information. In that case, information flows and communication channels need to be improved. More integrity in the public debate is required. Otherwise, powerful technologies that can contribute to food security and sustainable development will remain underutilized, leading to unnecessary human suffering and environmental damage.

The concluding chapter 8 provides a summary of the evidence so far. While the experience with impacts of already commercialized GM crops is predominantly positive, this experience is still limited to a few concrete examples. More interesting future GM crop applications may produce much bigger benefits. GM crops are not a magic bullet for agricultural development. They should not be seen as a substitute for other technologies and much needed institutional innovation. But GM crops can contribute to sustainable agricultural development, if the blockade of public resistance can be overcome. The outright rejection of GM crops by many overshadows other critical points that definitely need more attention, such as ensuring sufficient and equitable access to suitable seed technologies by poor farmers and avoiding increasing concentration in the crop biotech industry. Such issues cannot be solved by banning GM crops, but by enhancing the institutional and regulatory environment. Given the global challenges ahead, sustainable agricultural development and food security will not be possible without harnessing the potentials of plant biotechnology, including GM crops and other promising new techniques.

CHAPTER 2

PLANT BREEDING AND AGRICULTURAL DEVELOPMENT

Striving for sufficient food has always been at the heart of human existence. This chapter explores how humankind has evolved from scavenging to hunting and gathering and finally to producing food in a systematic manner. Plants have been at the center of this process as they, directly and indirectly, provide virtually all of our food. While initially almost all humans were inevitably involved in the sourcing of food, the start of agriculture made it possible for people to pursue other occupations, marking the beginning of civilization. Given the subsequent explosion of the world population, it has always been the prime objective of agriculture to increase the supply of food. I outline the fundamental advances in agricultural technology that have made sufficient food production growth possible in the past, also discussing related economic, social, and environmental implications.

A particular focus in this historical overview of agricultural development will be on plant breeding, which has always been one of the most important factors in increasing agricultural productivity. The systematic intervention by humans in natural selection of plants began with agricultural cultivation around 12,000 years ago and has since developed into a complex science. Nevertheless, even as plant breeding became more systematic over time, the improvement of crop varieties has always relied on the same underlying concepts. This development will be summarized to clarify that genetic engineering is not a drastic change of principles, but part of a historical continuum of advancements to further increase the efficiency of crop improvement required to face the global challenges ahead.

The Beginnings of Agriculture

Human history stretches over several million years, an evolutionary period that was accompanied by various shifts in the way food was obtained. For a long time, humans were dependent on scavenging dead or trapped animals, making the gathering of edible plants an important complement of securing a stable supply of food. Even when hunting was added, the gathering of plants remained essential for a balanced human diet. The nomadic hunter-gatherer way of life persisted for millions of years until it was succeeded by sedentary farming, featuring the organized cultivation of plants, about 12,000 years ago. The eventual transition from foragers to farmers, a process known as the Neolithic Revolution, has been investigated by various scholars, leading to different notions about where and why humans started to cultivate plants and later domesticate animals (Mannion, 1995; Barker, 2006).

There seems to be consensus that the origins of farming date back to around 12,000 years ago. At that time, global temperatures began to rise, marking the end of the Ice Ages (Pleistocene) and the transition to the current warm period (Holocene). A widely held view is that several villages in the Levant (Eastern Mediterranean) began to cultivate emmer and einkorn, marking the origins of agriculture. Similarly, other regions in the Fertile Crescent (Mesopotamia and Nile Valley of Northeast Africa), are thought to have cultivated other cereals and pulses not long after. Yet, archeologists found that the potato was first cultivated in the Peruvian Andes, probably even earlier than einkorn and emmer in the Levant. Consequently, there are theories arguing that the domestication of plants began independently in different parts of the world and at different times.

The Swiss botanist Augustin de Candolle suggested that crops must have been domesticated in areas where their wild relatives can be found. The Russian geneticist Nikolai Vavilov subsequently discovered that there are several geographically separated regions across the globe that are endowed with vast genetic variability of various crop species. Vavilov concluded that cultivation of the respective crops must have started in these "Centers of Origin," which are found in the Andes, Mesoamerica, the Fertile Crescent, China, India, and Ethiopia (Murphy, 2007a). The American geneticist Jack R. Harlan augmented this concept, suggesting a system of only three centers, from which plant material was transferred to nearby "noncenters" (Harlan, 1971). Vavilov's concept of Centers of Origin remains a popular way of depicting that the domestication of crops occurred independently in different parts of the world and at different times.

The reasons why humans eventually shifted from collecting plants to cultivating them are not completely clear. It is unlikely that the initial move to cultivating plants was a targeted decision because no form of agriculture had existed previously, so there was no ideal to strive for. Rather, farming must have evolved as a consequence of particular factors. A popular view is that the change in climatic conditions, which caused the end of the Ice Age, shifted the resource base, concentrated people, plants, and animals in oases, and provided, for the first time, a sufficiently warm and moist climate to allow for the cultivation of crops. Alternatively, it has been suggested that sedentary life in combination with an increasing population may have led to the depletion of wild resources, forcing people to innovate, which ultimately resulted in cultivation and domestication. The spare time created through sedentary life may also have facilitated experimentation and innovation. Yet another explanation could be increasing awareness of risk among humans, resulting in a desire to produce and store a surplus of food in case of future shortages (Mannion, 1995; Diamond, 1999). It is likely that all these factors and possibly others had played a certain role.

Over the course of the following millennia, starting about 10,000 BC, hunting and gathering was gradually replaced by various farming systems all over the world. The Neolithic Revolution is considered one of the most path-breaking events in human history, as it laid the foundations for further social, cultural, and economic development. The most fundamental effect was that agriculture allowed the same land to support much more people than was possible under hunting and gathering. This enabled humans to produce more food than they required for subsistence, creating capacity for barter and exchange and enabling others to specialize on non-food producing occupations. Furthermore, products for shelter, heating, clothing, bedding, animal feed, and many other uses could be produced in abundance. The agricultural revolution was thus a precondition for the establishment of the first urban civilizations in Egypt, Mesopotamia, and the Indus Valley (Diamond, 1999; Barker, 2006).

The Race between Food Production and Population Growth

Evidence from early farming societies shows that crop cultivation and population growth were closely correlated. However, determining with certainty which was the cause and which the effect is an impossible task (Diamond, 1999). On the one hand, population pressure may have depleted wild resources to such an extent that hunting and gathering could no longer feed the population sustainably, forcing people to find ways of producing food. On the other hand, as explained, crop cultivation may have started due to other reasons and subsequently may have enabled population growth in the first place. Moreover, sedentary life, which was encouraged through farming, allowed families to have children in shorter time intervals. Nomadic hunter-gatherers are required to wait until a given child is old enough to walk with the tribe before having a new baby. This is not the case for sedentary farmers, providing another possible reason to believe that population growth followed the rise of agriculture.

The complex relationship between population growth, food supply, and economic wealth also resulted in contrasting theories of human development. In his famous *Essay on the Principle of Population*, Thomas Robert Malthus postulated in the late eighteenth century that population growth would inevitably lead to widespread famine and disease. Malthus built his argumentation on assumptions that the population would grow exponentially, whereas food supply could only grow linearly. As a result, food supply would naturally keep population growth in check. The Malthusian catastrophe did not materialize, mainly because the assumption of linear food supply growth did not take into account the possibility of technological improvements to increase agricultural yields. Malthus considered the expansion of land to be the only option for increasing food supply. Furthermore, he did not consider that birth control could start to reduce population growth at some point. In contrast to Malthus, Ester Boserup argued in the opposite direction. In her book, *Conditions of Agricultural Growth: The Economics of Agrarian Change under Population Pressure*, which was published in the mid-1960s, Boserup argued that population growth would stimulate technological development. According to her, rising demand for food would require farmers to increase the productivity of their land, forcing them to innovate and intensify production, thus driving agricultural progress. Boserup was convinced that humanity would always find a way to increase food production through innovation, if need arises.

While the question whether—in early times—agricultural production contributed to population growth or vice versa is of historical interest, it is of lesser relevance for agricultural development in modern times. Nowadays, growth in agricultural production and food availability do not cause further population growth. On the contrary, the demographic transition observed around the world shows that human fertility rates decline significantly with rising incomes and improvements in nutrition and health. Nevertheless, population growth remains high in poor regions of the world, so that the question how to feed the growing number of people remains as relevant today as it has always been in the history of agricultural development.

Agricultural Technology and Intensification

The transition from gathering to producing food was not achieved over night. The early farming societies went through a process of trial and error that was likely interrupted by periods during which they reverted back to gathering. Also the distinction between nomadic foragers and early sedentary farmers was not as clear-cut as one might expect. Many early farmers continued to move around, while many foragers settled down, at least for certain periods of time. As food production gradually became more successful, a growing number of gatherers began to imitate it. This led to a millennia-long process of spreading farming systems throughout the world. In Central Europe and the Americas, agriculture only became more popular between 6,000 and 3,500 BC (Diamond, 1999).

During this expansion process, early farmers around the world naturally began to seek ways of improving the effectiveness of their work. Due to their lack of knowledge regarding plant reproduction, early farmers had only very limited means of influencing the domestication of plants in an intentional way. Instead, they concentrated on improving the efficiency of farming operations, such as seeding and harvesting, and storage facilities to reduce post-harvest losses (Murphy, 2007a). The domestication of animals was another important step for generating more food directly and indirectly. Animals did not only provide milk and meat, but they also produced manure, which could be used as fertilizer. Moreover, animals were capable of transporting heavy weights and pull ploughs, after these had been invented, allowing the cultivation of previously unworkable soils (Diamond, 1999). Ploughs were invented in the Fertile Crescent around 4,500 BC, initially made of stone. Metal ploughs were invented around 1,200 BC. The first irrigation systems, in the form of extensive canal networks that ran through the fields, were developed between 6,000 and 5,000 BC in Ethiopia and the Nile Valley. The first farming manuals were written by the Babylonians around 1,700 BC, recording agricultural practices such as crop rotation systems (Murphy, 2007a).

Many of these inventions did not reach Western and Central Europe until the Middle Ages. Instead of trying to increase productivity, farmers in Europe simply expanded the cultivated area or moved to more fertile land in order to maintain a sufficient food supply as the population grew (Murphy, 2007a). Until the end of the fifteenth century, there was not much home-grown agricultural innovation in Europe. Some inventions were adopted from the Islamic world. This trend was only reversed about 500 years ago. Since then, European countries have been major drivers of agricultural modernization, a process that is closely linked to the industrial and scientific revolutions that followed later. The same holds true for North America.

European agronomists contributed improvements such as the advanced crop rotation systems of Richard Weston, who propagated the use of break crops during fallow periods in 1645, as well as mechanical technologies, such as Jethro Tull's seed drill (from 1701) and Andrew Meikle's thresh- ing machine (from the 1780s). European countries have also significantly shaped agricultural progress in terms of new knowledge in chemistry and biology. Since the rise of agriculture it has always been a pressing con- cern to minimize the various risks that crops are exposed to, including competition from weeds and harm from insect pests and diseases. Early farmers had already begun to experiment with various substances avail- able to them, including chalk, alum, and sulfur, to protect their crops. These efforts became increasingly educated and systematic from the sev- enteenth century onward. In Europe, several dozen chemicals for crop protection were widely used as early as 1850, a notable example being the "Bordeaux mixture," consisting of copper sulfate and hydrated lime, for weed control in French vineyards (Murphy, 2007a). The twentieth century saw the development of an agrochemical industry to produce and commercialize a large number of crop protection pesticides.

Similar developments occurred in fertilization. Plants obtain the nutrients they require from the soil on which they grow, so—in order to continuously cultivate crops on a given piece of land—it is essential that these nutrients are replenished on a regular basis. Nitrogen, one of the most important plant nutrients, was originally obtained from biological sources such as animal manure. However, the availability and effective- ness of such organic fertilizers remained limited. In the 1840s, Justus von Liebig discovered inorganic nitrogenous fertilizers that could be used to maintain or increase soil fertility more effectively. This led to the devel- opment of the Haber-Bosch process for fixing nitrogen gas into ammonia at an industrial scale and the subsequent formation of a chemical fertilizer industry. Around the same time, an inorganic form of phosphate, another vital plant nutrient, was discovered. The discovery of chemical fertil- izers has vastly increased agricultural yields, contributing immensely to feeding the world's growing population and reducing pressure to expand agricultural land.

The Beginnings of Plant Breeding

Several means of increasing food production have been mentioned: expanding the cropland, improving agronomic methods, mechaniza- tion, as well as the use of chemical fertilizers and pesticides. However, the single most important factor in improving agricultural productivity has been the genetic improvement of the crop plants themselves (Duvick, 1986).

Domestication refers to the irreversible genetic modification of organisms, in this case plants, to the extent that they would no longer survive in the wild (Blumler and Byrne, 1991). In other words, domesticated plants become more and more dependent on human intervention. Domestication is a direct result of human's natural desire to seek and retain the best specimens of a given selection of plants and seeds. Hence, domestication and cultivation of plants are inextricably linked. Wild cereals, for example, tend to shed their seeds in order for them to be dispersed by wind or animals. By cultivating plants and harvesting their ears of grain, early farmers unintentionally exerted enormous selection pressure by discriminating against those plants that shed their seeds before they were harvested. Any seeds dispersed before the harvest would simply fall to the ground and not be harvested. Thus, only plants that did not shed their seeds were given the chance to pass on their genes to the next generation. Eventually, the seed-shedding trait was lost entirely from the cultivated population. The result was a new phenotype with a superior trait from farmers' perspectives, but extremely maladapted to life in the wild. It should be noted that the original seed-shedding trait continues to exist in the wild relatives of the cultivated species. This process of unconscious selection also applies to many other traits, including the loss of seed dormancy, synchronous flowering, thin seed coats, and an upright posture (Murphy, 2007a; Hainzelin, 2013).

Based on this initially unconscious selection, farmers in ancient societies eventually began to consciously and systematically select plants according to observable characteristics, such as larger and more numerous seeds. Farmers started to only save the largest and best-looking seeds, or only seeds from the best-performing fields, for planting in the following year. The degree to which plants responded to selection pressure was likely genetically predetermined and varied among different species and types. The most favored types were those that were most responsive, for instance, by germinating soon after planting and producing larger seeds that did not shed from the parent plant. This encouraged further selection and more widespread use of the respective types. This form of intentional selection was still entirely based on visual observation, not on more profound scientific knowledge. Nevertheless, this form of selection resembles what modern plant breeders do and is thus often referred to as "pre-scientific empirical breeding" (Murphy, 2007b). This early form of plant breeding achieved significant results. Many new varieties emerged, which not only produced higher yields but were also adapted to a vast range of environments, as cultivation spread throughout the world. Moreover, it was gradually realized that not only observable characteristics, such as grain size and quantity, could be influenced by selection but also traits such as robustness against pests and diseases, taste, suitability for baking, and many more. The selection process eventually produced landraces, which are relatively stable varieties specific to certain regions and environmental conditions. Whether intentional or not, it is important to note that evolutionary change was not caused directly by humans. Humans merely contributed to influencing the environmental conditions to which organisms were exposed. The adaption to these changes is the natural process called survival of the fittest (Darwin, 1876).

The process of improving crop plants by selection was not only slow but also limited because only existing varieties could be used. Moreover, there was always the risk of events such as wars and pests wiping out entire populations of carefully selected plants. The seventeenth century saw the emergence of new scientific knowledge not only in agronomy, mechanization, fertilization, and crop protection but also in genetics. The German botanist and physician Rudolf Camerarius first demonstrated that plants reproduce sexually and, in 1694, suggested that pollen acts as the "equivalent of animal sperm in plant fertilization" (Murphy, 2007a, p. 257). Camerarius further hypothesized that crosses between different varieties could lead to new and superior crop varieties. The British botanist Thomas Fairchild confirmed this hypothesis when he developed the first human-made interspecific hybrid in 1718. These discoveries implied that plant breeding was no longer limited to only improving existing varieties through selection, but that new variation could be generated through crossing. This expanded the possibilities of plant breeding significantly.

However, the genetic mechanisms of heredity were not understood until 1900, when the work of Gregor Mendel, an Austrian monk, was rediscovered. Mendel's original contribution was published in 1865, but remained unnoticed for several decades. In fact, there were several botanists and biologists who investigated the subject of heredity in plant breeding in the second half of the nineteenth century. Mendel's work was the most systematic one. Mendel's basic hypothesis was that each characteristic of a plant is determined by two hereditary elements, one from each parent. To test this hypothesis, he carried out detailed experiments with 22 different varieties of garden pea to test seven observable traits (Mayr, 1982). In 1865, Mendel published three laws of inheritance, confirming his initial hypothesis.

While Mendel used somewhat different terminology than geneticists today, he correctly established that genes determine particular biological traits, such as seed color. Alleles, alternative forms of genes, determine the phenotypic expression of a given trait, for example, whether the seed color will be green or yellow. The first of his laws, the law of segregation, describes how pairs of alleles in the parents split during the formation of sperm and eggs (gametes), so that any sperm or egg only carries one allele for each inherited trait. When the sperm and egg unite during fertilization, the offspring has again a pair of alleles. The two alleles in the offspring do not blend but remain separable in order to be segregated again during the formation of future gametes. The second law, the law of independent assortment, simply states that this segregation of allele pairs occurs independently when more than one trait is considered. The third law, the law of dominance, proposes that there are dominant and recessive alleles. Organisms that have two identical alleles for a given gene (homozygotes) will always express the phenotype. In heterozygotes, organisms with different alleles for a given gene, only one of the alleles determines the organism's phenotype for that gene. This allele is called dominant, whereas the other allele, with no observable influence on the phenotype, is referred to as the recessive allele (Acquaah, 2012). In Mendel's experiments, the yellow seed color was, for instance, dominant over the green seed color.

Mendel's findings laid the foundations for further research on inheritance in the twentieth century. It was subsequently found that there are many exceptions to Mendel's laws. In certain cases, the explanatory powers of Mendelian inheritance are insufficient, something Mendel himself was aware of and which he pointed out at the time of publication. For instance, many traits are a result of the interaction between several genes. Such polygenic traits often show a wide range of phenotypes and can thus not be explained by simple Mendelian inheritance. Given the increasing number of exceptions, Mendel's laws have lost some of their usefulness and have been replaced by new or revised postulates. Nevertheless, Mendel's discoveries had a profound educational effect, encouraging further research in the field of genetics (Mayr, 1982). Arguably, the rediscovery of Mendel's work in 1900 eventually turned plant breeding into a much more knowledge-based science.

Modern Plant Breeding

Plant breeding can be thought of as the process of altering a plant's genotype in order to obtain a desired phenotype, with the aim of developing an ever more diverse range of superior plant varieties (Hainzelin, 2013). In a simplified way, all that is required for breeding a new plant variety is a certain degree of genetic variation within a given population and a way to identify and select the most suitable variants, making variation and selection the backbone of plant breeding (Murphy, 2007b).

Agriculture, in its simplest as well as its most modern form, requires crop varieties to display a certain degree of homogeneity in regard to traits such as time of germination and maturity, fruit size and quality, plant height, and many more. Consumer preferences and quality standards continuously reinforce this need by demanding new features. As a result, one of the main goals of plant breeding has always been to achieve genetic uniformity within a population, so that each individual of a given variety expresses its desired traits reliably. Populations can be thought of as large groups of interbreeding individuals, which may consist of thousands of individuals. The genetic variation among these individuals will produce a vast range of phenotypes. Depending on the means of reproduction of the species that makes up a given population, different selection methods are available to a plant breeder. Distinctions are made between self-pollinated, cross-pollinated, and asexually reproducing species. Important examples of these include wheat, maize, and potatoes, respectively.

The oldest and most basic form of selection is known as mass selection. Based on phenotypic expression, it consists of either selecting desirable individuals in a population (positive selection) or eliminating undesirable individuals from a population (negative selection), with the aim of increasing the frequency of the desirable genes. It is thus almost identical to the intentional selection practiced by the very first farmers. This method can be used to maintain the purity of existing varieties, adapt varieties to new environmental conditions, or develop entirely new varieties. It therefore serves as the basis for many other selection methods that are available to modern plant breeders. Recurrent (mass) selection, for example, involves crossing the selected individuals, which display the desirable trait, with one another in order to form a new population from which the best individuals are then again selected. Repeating this process will eventually produce a new variety that displays the desired trait reliably. Ideally, the resulting variety is equally rich in genetic variation as the original population, as this will make it responsive to further selection, enabling it to be used as a parent in a new cross. Similarly, the original population may be sorted into genetically pure lines by means of repeated inbreeding, or selfing, following the initial selection from the population. Inbred purelines are an essential input to hybrid breeding (see below).

Another very important breeding method, which was developed in the 1920s, is backcrossing. This entails crossing variety A, which is, for example, susceptible to a certain disease but otherwise desirable, with variety B, which is resistant to the disease but otherwise inferior. The offspring is then selected according to presence of the resistance and crossed again (backcrossed) with the overall superior variety A. This process can be repeated until the offspring uniformly displays the disease resistance of variety B in addition to all the desirable characteristics of the initial variety A. Rather than trying to find overall superior varieties, this allows plant breeders to look for specific traits to be added to certain varieties, which has made old landraces yet again a valuable resource. While these methods were originally based on phenotypic selection, it is nowadays possible to select at the genotypic level, which, in some cases, has significantly sped up the breeding process. The underlying concept, however, has always remained the same (Acquaah, 2012).

Paradoxically, the strive for homogeneity destroys the very foundation in breeding—genetic variation. Millennia of selection have taken their toll on naturally occurring genetic variability. Traditional, relatively non-uniform landraces, which represent the very first domesticated crop varieties, are becoming increasingly rare (Murphy, 2007b). On the other hand, plant breeding has also produced tens of thousands of new plant varieties over the past millennia. And, not all landraces were lost; many merely disappeared from agricultural cultivation but were preserved in crop-breeding centers and gene banks. Nevertheless, the question arises of how today's plant breeders obtain the genetic variation required for their breeding programs.

In order to breed plants sustainably, it is necessary to find ways of regaining the variation that is inevitably lost in the process of creating new, genetically uniform varieties. While new variation is sometimes created naturally by spontaneous mutation or hybridization, such events are rare and cannot be relied upon exclusively. One way of actively introducing new variation is to import exotic plant varieties from other regions, in order to cross them with local varieties. For this purpose, many Western European countries sent exploration voyages around the world to collect landraces as early as the sixteenth century. Although these expeditions became more systematic in the eighteenth and nineteenth centuries, a famous example being Charles Darwin's voyage on the HMS Beagle from 1831 to 1836, there are obvious limits to the effectiveness of this method, as well (Murphy, 2007a).

With the emergence of scientific plant breeding, new ways of inducing variation were discovered, most notably hybridization and induced mutagenesis. Hybridization simply means crossbreeding two individuals. In nature, this usually occurs between individuals of the same species (intraspecific), however, so-called wide crosses between varieties of different species (interspecific) or even different genera (intergenus) are also possible and have, in fact, been instrumental in creating variation in many of today's major crops. For example, today's rapeseed (canola) is the result of a spontaneous cross between two species of the *Brassica* genus, namely cabbage and turnip, which occurred about 2000 years ago (Murphy, 2007a). Intergenus hybridization is rare, but it does occur spontaneously and has produced a few highly useful species, such as durum wheat and breadwheat. In much the same way, plant breeders try to make wide crosses deliberately, to create variation that can subsequently be used to develop genetically stable varieties. As wide crosses lead to the exchange of genes across species boundaries, the resulting varieties are "transgenic," although they are not considered as such from a regulatory perspective, as will be discussed in chapter 6.

Induced mutagenesis is a different approach that exploits mutations. A mutation is a sudden genetic change in an organism, which may ultimately alter its phenotype much more radically than hybridization. All living organisms are constantly exposed to all kinds of mutagenic agents such as sunlight and numerous naturally occurring as well as human-made chemicals. Mutations were discovered as a natural source of variation, but, as not all mutations are inheritable, the likelihood of a useful spontaneous mutation, in the sense of improving a given crop variety, is very low. In order to increase the occurrence of mutations, scientists tried to deliberately induce them under experimental conditions as early as 1901. Around 1930 it was discovered that X-rays and chemicals could be used to vastly increase the occurrence of gene mutations (Mayr, 1982). Plants created using mutagenesis are sometimes called mutagenic plants, and the resulting varieties are referred to as mutant varieties. The FAO/IAEA Mutant Variety Database reported that in 2014 there were over 1,000 mutant varieties of major staple crops grown on tens of millions of hectares around the world.

Hybrids and the Launch of the Seed Industry

With the rapid progress in plant breeding over the last 100 years, new regulatory, economic, and social issues arose. New knowledge can spread relatively easily. While this can be very useful and has indeed been instrumental for development, when knowledge is of high value transferability can become a problem. One of many examples of such valuable knowledge is information that relates to the invention of new products or processes that can be copied or imitated at relatively low cost by customers or competitors. In such cases, there is no guarantee that economic returns will be sufficient to compensate the inventor for the time, money, and effort invested in R&D. For private inventors that depend on revenue and profit, such conditions may not provide sufficient incentives to invest in R&D, a problem known as lack of appropriability. Patent systems since 1623 have attempted to counteract this problem by granting inventors a certain degree of market power for a limited period of time, enabling them to extract adequate returns from their investments (Cohen and Levin, 1989). However, until recently patents could not be obtained for biological inventions such as plant varieties.

With the start of a more scientific approach to plant breeding in the early twentieth century, the resources required for running a breeding program and the types of actors involved changed. Farmer plant breeders, who were engaged in simple selection and accounted for the majority of the world's breeding progress until then, made way for professional plant breeders. An important part of agricultural modernization was therefore the separation of crop cultivation and seed production (Hainzelin, 2013). Initially, scientific plant breeding was a matter of the public domain. Although the very first seed companies were established already in the eighteenth century (FAO, 2009), these early companies mainly multiplied and sold seeds of existing varieties rather than having their own breeding program. Many of the early seed companies were family owned, having emerged from farming households, and lacked the resources to fund significant R&D activities. Moreover, many of the early seed companies were concerned with vegetable seeds, leaving growers of field crops almost exclusively dependent on seed saved from their own harvest or obtained from trading with neighboring farmers (Fernandez-Cornejo, 2004).

Toward the end of the nineteenth century, the first national seed associations were established in Europe and America. Their purpose was to provide quality assurance to farmers in the form of seed certification programs in an attempt to encourage commercialization of seeds. Despite these efforts, private interests in plant breeding remained limited. The main reason is the natural reproducibility of seeds or the fact that grain, the product of agricultural cultivation, can also be used as seed, the main input to agricultural cultivation. Thus, farmers can simply save a fraction of the harvest and use it as seed in the following year, with minimal loss in yield and quality. Although this reproducibility allowed farmers to cultivate crops in the first place, the varieties used in ancient times had evolved naturally and their development was not associated with high costs. Nowadays, the development of a new crop variety is costly and can take ten years or more. Nonetheless, the same natural reproducibility of seeds remains intact. Once a farmer has purchased seed of an improved variety, the product can be replicated almost infinitely at comparably low costs. This strongly differentiates seed of improved plant varieties from other innovative products (Kloppenburg, 2004).

As mentioned, patent laws or other intellectual property rights (IPRs), which apply to technologies in other industries, were not available for plant varieties until recently. An internationally recognized framework for the protection of plant breeders' rights was only established in 1961 under the International Union for the Protection of New Varieties of Plants (UPOV). The UPOV conventions were updated several times since then, but plant variety protection under these conventions remains weaker than protection through patents. Limited appropriability was the main reason for the relatively low interest of the private sector to invest in breeding of major food crops.

In the 1930s, public plant breeding efforts in the United States developed so-called hybrid maize varieties, obtained from crossing two highly inbred parent lines. Such crosses generate genetically homogenous offspring, which produce yields in excess of either parent line, an effect known as heterosis or hybrid vigor. Hybrids are differentiated from so-called open-pollinated varieties (OPVs) that were not developed through hybridization and have low levels of heterosis (this also includes line varieties for self-pollinating crops). It should be noted that the word hybrid is sometimes also used to refer to any offspring of two different species or varieties. However, in a breeding context the word hybrid is mostly used to describe varieties that display a high degree of heterosis. The adoption of hybrid maize by US farmers was unprecedented in the history of agricultural innovations. By 1965, over 95 percent of the total maize area in the United States was planted to hybrid varieties. Maize yields, which had begun to decline in the early twentieth century, increased rapidly from the 1930s onward, relieving, once again, pressure on food security and agricultural land conversion. Between 1930 and 1965, maize production increased by almost 60 million tons while land devoted to maize production decreased by over 12 million hectares. The development of hybrid maize varieties is therefore often considered one of the most important achievements of agricultural science (Kloppenburg, 2004).

Another important feature inherent to hybrid seed is that it only displays its hybrid vigor in the first generation (the F1 generation). Replanting the grains from this generation yields significant genetic variation in the F2 generation, leading to a heterogeneous and lower-yielding crop. It follows that, for farmers who wished to benefit from the hybrid vigor of the new maize varieties, it was no longer economical to save and replant seed. Instead, farmers began to purchase fresh seed on an annual basis.

The development of hybrid technology affected the interest of private companies in maize breeding considerably. Firstly, the large yield gains that hybrid maize varieties could produce increased the farmers' willingness to pay for seed. Secondly, for the first time in the long history of plant breeding, the development of superior crop varieties became an economically viable prospect. Hybrids changed the reproducibility of seeds in spite of lacking IPR protection. The newly gained appropriability has caused a substantial increase in private investments into maize breeding since the 1930s. Most existing seed companies quickly expanded and diversified into the development of hybrid maize varieties, and many new seed companies were formed for the same purpose (Fernandez-Cornejo, 2004).

For some time, maize remained the only crop for which hybrid varieties had been developed. Breeding of non-hybrid crops, which consequently included all other field crops, remained unprofitable and was limited to efforts by the public sector. Nevertheless, hybridization of maize initiated the commercialization of seeds, leading to rapid progress in plant breeding, fuelled by private sector investments. Nowadays, hybrid technology is also available for several other crop species, including millet, sorghum, rice, wheat, and cotton. However, due to technical constraints, hybridization costs are still high in some of these crops, so that hybrid varieties are not grown as widely as hybrid maize.

The Green Revolution

Many of the agricultural advancements of the first half of the twentieth century were concentrated in North America and Europe. However, rapid population growth and increasing severity of food shortages in the mid-twentieth century began to raise awareness of the need to further increase agricultural productivity, especially in developing countries. Efforts by various bilateral and multilateral agencies to provide agricultural assistance to countries in Asia and Latin America were of limited success because the modern agricultural technologies from temperate industrialized countries were often of little use in the tropical conditions of many developing countries (Borlaug, 2000). In light of this, the Mexican government and the Rockefeller Foundation started to collaborate in 1943 to establish a research institution in Mexico that later became the International Maize and Wheat Improvement Center (CIMMYT). The goal of this new institution was to improve local varieties of important food crops to promote food security in Mexico and other developing countries.

The American Scientist Norman Borlaug was appointed as chief wheat breeder of the Mexico-Rockefeller Program. Borlaug crossed highly productive North American wheat with varieties from developing countries that were adapted to various tropical conditions. Almost two decades of recurrent selection finally produced high-yielding varieties (HYVs) of wheat for use in tropical and subtropical climates. These HYVs were much more responsive to nitrogen fertilizers, thus producing higher yields per unit of area and time, given sufficient nutrient and water availability. In order to prevent plants from collapsing under the increasingly large ears of grain, they were also "dwarfed," meaning that their stalks were strengthened and reduced in height. Both the high-yield and dwarf characteristics were achieved by systematic crossbreeding followed by careful selection (Pearse, 1980).

Once these improvements were beginning to take effect, the next breeding objectives were to enhance pest and disease resistances, which were achieved in a second phase (Hainzelin, 2013). Simultaneously, several agricultural research centers with similar goals were established in other parts of the world, most notably the International Rice Research Institute (IRRI) in the Philippines, which aimed to produce HYVs of rice. In 1971, four international agricultural research centers were combined under the umbrella of the Consultative Group on International Agricultural Research (CGIAR). Today, the CGIAR Consortium comprises 15 such international research centers.

The development of HYVs was accompanied by the implementation of widespread rural development programs in many countries of Latin America and Asia. These programs included agricultural input subsidies, access to credit, access to output markets, import protection, and agricultural extension. The overarching aim was enabling smallholder farmers to adopt agricultural innovations, such as HYVs, irrigation, chemical fertilizers, and crop protection measures (Hainzelin, 2013). The development and widespread adoption of HYVs in developing countries since the 1960s—coupled with the higher use of other agricultural inputs—is often referred to as the Green Revolution. It is important to note that the Green Revolution varieties of wheat and rice were all OPVs, so that farmers could reproduce their own seeds once they had obtained HYVs through formal or informal seed channels.

The economic and social impacts of the Green Revolution varied significantly between regions. The HYVs were dependent on the availability of water and agrochemicals, so the advantages were especially large under favorable agroecological and infrastructure conditions. Thus, while wheat yields in Mexico and wheat and rice yields in India and Pakistan nearly doubled between 1965 and 1970, HYV adoption remained much lower in sub-Saharan Africa (Evenson and Gollin, 2003). Even today, irrigation and road infrastructure is much less developed in Africa than in large parts of Asia (Juma, 2011). It was also observed that relatively larger farms adopted the new technologies first, while smallholders followed with a certain time lag, due to higher risk aversion and more severe institutional constraints.

Nevertheless, the increase in agricultural production was instrumental in keeping grain prices relatively low, benefitting consumers of food (Hainzelin, 2013). Many smallholder farm households are also net buyers of food, that is, they buy more food than they sell in the market. And, while regional inequality between farmers in favorable and marginal regions may have increased, studies show that the Green Revolution contributed significantly to rural poverty reduction through higher farm incomes and employment generation (Hazell and Ramasamy, 1991). Norman Borlaug, who is considered the "father of the Green Revolution," received the Nobel Peace Prize in 1970 for his achievements in plant breeding, which are credited with preventing the deaths of millions of people in the developing world.

However, the Green Revolution has also been associated with environmental problems, such as the overuse of chemical fertilizers and pesticides, which has sparked widespread controversy (Eicher and Staatz, 1998). Misuse of agrochemicals may not only harm the plants to which they are applied but also incur substantial adverse effects on the environment in the form of soil degradation and pollution. After such misuse was observed during the 1960s and 1970s, further research has tried to reduce negative environmental externalities through improving agronomic practices, breeding more robust varieties, and increasing the selectiveness of agrochemicals. Nevertheless, in some regions agrochemicals remain overused. The homogenization of cultivated varieties, a result of HYVs replacing less profitable landraces, has also contributed to reduced varietal diversity (Brush, 2000). And the research focus on only a few cereals has reduced the diversity of cultivated species. Of the 50,000 plant species that could, in theory, be used to feed humans, only 15 provide about 95 percent of all calories consumed worldwide. Wheat, maize, and rice alone account for almost 60 percent of the calories consumed (Becker, 2011).

This loss in agrobiodiversity is no surprise; it is the collective consequence of economic decision-making by individual farmers deciding to grow the most profitable crops and varieties. Low food diversity is undesirable from a nutritional quality perspective. Agrobiodiversity erosion may also reduce the resilience of agricultural systems, which is considered problematic from a socio-ecological perspective. On the other hand, it must not be forgotten that the specialization on the best performing species and varieties has vastly increased the productivity of agricultural land. This has reduced the need for agricultural expansion, protecting vulnerable areas from being converted to cropland

and allowing biodiversity to prosper in natural ecosystems (Lipton and Longhurst, 1989; Hainzelin, 2013).

Advent of Molecular Tools

The development of improved crop varieties through conventional breeding is a relatively laborious and time-consuming process. Typically, it takes 10–15 years from the first cross to the commercial release of a new variety. This also involves a high cost for the breeder. Hence, technologies that are capable of reducing the cost and time required and increasing the likelihood of success in developing varieties with improved traits are welcome in principle. With this objective in mind, several molecular tools were developed and used in plant breeding over the last few decades. These new tools have not altered the general concept of breeding new varieties, consisting of generating initial genetic variation and selecting the best performing candidates for further testing in the field. But various tools of biotechnology have contributed to increasing the efficiency of breeding considerably.

Biotechnological tools that are used in breeding include cell and tissue culture techniques, allowing the *in vitro* regeneration of whole plants from isolated plant parts, such as leaf tissue, or from individual cells. Cell and tissue culture techniques are used, for instance, for the rapid multiplication of plants with desirable traits. A different tool that has been widely used in breeding since the early 1990s is marker-assisted selection (MAS), sometimes also referred to as SMART breeding. Molecular markers are fragments of DNA (deoxyribonucleic acid) that are linked to and jointly inherited with plant traits of interest. Unlike the traits themselves, the markers can easily be detected with molecular methods. Thus, the process of selection for desirable traits can be based on the genotypic information, which can be much faster than waiting until the trait is phenotypically expressed in the growing plant. MAS can be especially useful to select for traits that are difficult to measure, have low heritability, or are expressed late in plant development (Acquaah, 2012).

Yet another biotechnological tool is called protoplast fusion. This involves the isolation of cells from two distinct organisms and the removal of the cell walls to produce protoplasts. The two protoplasts are then fused using chemical treatment or electric shocks, resulting in a nucleus that contains the genetic information from both organisms. Protoplast fusion allows the combination of genes across species boundaries, hence the resulting varieties are "transgenic." Nevertheless, protoplast fusion does not count as genetic engineering because it does not involve a direct intervention into the structure of the genetic code. Genetic engineering is described in more detail in chapter 3.

Impact of Plant Breeding

The transformation from wild plant to cultivated crop of virtually all of today's crops occurred before the age of science and is attributable to simple recurrent selection by farmers. Perhaps one of the most impressive results of pre-scientific breeding is maize. Thanks to the study of genetics, the story of maize can be traced back to the wild grass teosinte, which, despite of looking very different from today's maize, is genetically very similar to it. Initial selection of teosinte goes hand–in–hand with the origins of crop cultivation in what is now Mexico over 10,000 years ago. The success of this early selection process is demonstrated by cave findings from Central America, which show that the wild grass teosinte was turned into something that closely resembles today's maize within a few millennia. The development of maize as a domesticated crop is thus not the result of modern scientific plant breeding but of millennia of selection by Central American aborigines. Consequently, when maize was first brought to Europe in the fifteenth century, it was already very similar in size and shape to today's maize (Becker, 2011).

Cereals such as wheat and rye are among the oldest crops. They are descendants from some of the first wild grasses that were domesticated in the Fertile Crescent at the beginning of the Neolithic Revolution. As was explained earlier, some of the first characteristics to be selected in the early landraces of cereals were loss of the otherwise vital seed-shedding and seed dormancy traits, which allowed these varieties to undergo further genetic improvement. Potatoes were also cultivated from a very early age, leading to a similarly remarkable transformation. What is rarely known today is that most wild forms of potatoes contain bitter compounds that make them toxic. Edible potatoes that are free from bitter compounds are a result of spontaneously occurring genetic mutations, which were first discovered by the indigenous people of Peru. While today's potato varieties stem from these mutants, they have gone through extraordinary change since then regarding size, robustness, quality, and fields of use (Becker, 2011).

Even sugarbeet, a relatively new crop, saw the majority of its transformation into a domesticated crop before knowledge-based science entered plant breeding. Selection for size, shape, and—most importantly—sugar content in sugarbeet only began in the nineteenth century. Early plant breeders, merely employing simple selection, were able to triple sucrose contents to 15 percent in 1900. Since then, sugar contents have increased by another two to three percentage points, a negligible achievement compared to the increase in the pre-science era. This is not to say that scientific breeding has played an insignificant role in the improvement of sugarbeet and other crops. In many cases, improvements in one characteristic, obtained by initial selection, led to detrimental effects on other traits. For example, the increases in sugar content of sugarbeet that had been achieved by 1900 were accompanied by decreases in yield. It was scientific plant breeding that later remedied this by increasing beet size, improving numerous agronomic characteristics and introducing resistance to diseases (DFG, 2010).

In a similar way, plant breeding has produced major improvements in many other crops. Both, age-old selection and more knowledge-based breeding during the last 100 years have generated domesticated crops that look very different from their wild ancestors and have very different characteristics—changes that would never have occurred naturally without human intervention. For quite some time, breeders have tried—and often managed—to circumvent the rules of natural heredity, transferring genetic material across species boundaries and causing artificial mutations in plants. Hence, the notion that breeding was entirely natural before the advent of genetic engineering is simply incorrect.

Figure 2.1 shows the yield increases of some of the world's major cereals since the 1960s. For all crops shown, yields more than doubled during

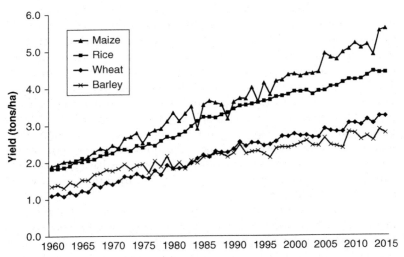

Figure 2.1 Worldwide yield developments of major cereals since the 1960s.
Source: Own presentation with data from FAO and USDA.

this period, wheat yields even tripled. Several factors have contributed to these gains in productivity, including the availability of affordable mineral nitrogen fertilizers, better pest and weed control, more widespread availability of irrigation, as well as improved agronomic practices and machinery. However, as mentioned before, the single most important factor has been the genetic enhancement of crop varieties, vastly improving agriculture's most essential input; the seed. Studies have estimated that—over the last few decades—plant breeding has been responsible for 50 percent of the yield increases in some of today's major field crops, including wheat, rice, maize, and soybean (Duvick, 1986; Evenson and Gollin, 2003). Whereas some of the largest yield increases occurred in agro-ecologically favorable environments, in many cases notable progress was also achieved under suboptimal conditions in regard to soil fertility, water availability, and pest and disease pressure (Eicher and Staatz, 1998; FAO, 2009). In spite of the rapid worldwide population growth over the last 50 years, the agricultural productivity gains led to improvements in per capita food availability, economic accessibility, and food security.

Current and Future Trends

The large increases in agricultural production resulting from the technological advancements have caused global food prices to decline significantly during the twentieth century. The long-term negative price trend made many believe that food prices would always decline further with economic development, also because the income elasticity of demand for food tends to be lower than for non-food consumption goods. However, with the start of the new millennium global food prices have not fallen further. Especially the price spikes in 2007–2008 and the following years have demonstrated that food scarcities are not only a phenomenon of the past (FAO, 2011). Extreme price peaks are usually of short duration, but a focus on the peaks alone masks the fact that the food price trend has been increasing during the last 15 years. Whether rising prices are good or bad depends on who is being asked. Farmers who are selling food are happy about rising revenues and profits, whereas consumers who purchase food can only afford less when prices soar. Smallholder farmers in developing countries are often both, sellers during some parts of the year and buyers during other times. The majority of the world's poor and food-insecure households buy more food than they sell in the market, so that they suffer from rising prices. Hence, price increases tend to aggravate hunger and food insecurity in the world (Ecker and Qaim, 2011; FAO, 2011).

Unlike the twentieth century, the last 10 to 15 years have seen growth in agricultural demand outpace growth in supply. What has happened?

While population growth rates are not as high anymore as they used to be in the past, the global population is still growing further with more than 70 million additional people added to the world every year. Furthermore, poverty reduction and income growth in many developing countries have contributed to rapid growth in the demand for animal source foods, especially meat and dairy products. Livestock production is often based on grain-based feed with significant conversion losses in food energy. Depending on the type of animal, the production of 1 kg of meat may require 5–10 kg of grain. In addition, agricultural products are increasingly used for non-food purposes such as biofuels, the importance of which has recently increased significantly. During the last 15 years, total demand for agricultural products—including food, feed, and other uses—has increased with an average of 1.8–2.0 percent per year (Qaim and Klümper, 2013). At the same time, growth in agricultural production has fallen. Figure 2.2 shows that the annual yield growth for cereals has fallen from 2 percent and higher until the 1980s to less than 1.5 percent today.

The trend in Figure 2.2 is aggregated over all major cereals. It should be noted that there are significant differences between cereal species. While yield growth for wheat and rice has recently been below 1 percent per year, maize yields are still growing with around 2 percent. The higher growth in maize may be due to the fact that hybridization technology is used much more in maize than in other crops, also generating larger

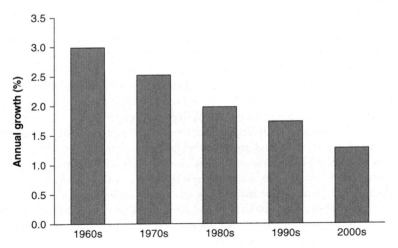

Figure 2.2   Worldwide yield growth in cereals since the 1960s.

Source: Own presentation with data from FAO.

incentives for the private sector to invest in maize breeding. Progress in other crops has relied much more on the public sector, but public sector agricultural R&D investments have declined in many parts of the world in the 1980s and 1990s (Eicher and Staatz, 1998). To some extent, the reductions in public R&D investments were due to the Green Revolution and other technological advancements in the 1960s and 1970s. Declining food prices had contributed to the notion that hunger is primarily a distribution problem and that agricultural technology would not have an important role to play for food security anymore. As a reaction to the changing price trend, public R&D investments were increased again more recently, but 20 years of relatively low investments are difficult to make up.

Independent of R&D investments, it is also important to stress that—in spite of the important role of breeding—a significant part of the yield growth in the last 50 years was due to higher amounts of mineral fertilizers and irrigation water being used. These types of yield increases have been exhausted in many parts of the world (Huang et al., 2002; Brown, 2012). Further productivity gains will have to come from other innovations, especially improved agronomy and seeds that can help to produce more with lower quantities of external inputs. Progress in breeding, however, is limited by the genetic variation that is found or can be generated in the species of interest. Conventional approaches alone will hardly suffice to maintain past rates of progress in breeding on a sustained basis. Genetic engineering offers large potential to increase the genetic variation that can be used by breeders to face the challenges ahead (Tester and Langridge, 2010).

Growth in agricultural demand will continue to be significant over the next few decades due to a further rising world population, rising incomes, and a growing use of plant-derived products and substances for energy and other industrial purposes (Borlaug, 2007; Godfray et al., 2010; Hertel, 2015). To prevent food prices from rising steeply, which would aggravate food insecurity, agricultural supply will also have to rise significantly, and at rates higher than those observed during the last 15 years. Increasing scarcities of natural resources, such as land and water, environmental problems, such as soil degradation and biodiversity loss, and changing climatic conditions pose additional challenges that need to be addressed with improved technologies. It is sometimes argued that—instead of increasing production—greater priority should be on reducing post-harvest losses, waste, and curbing the consumption of meat. Unquestionably, reducing losses and waste are important objectives, and, especially in rich countries reductions in meat consumption would also be desirable. However, such changes are not easy to achieve, and alone they would not suffice to reverse the global trends in demand and supply (Qaim and Klümper, 2013). Hence, various strategies need to be pursued simultaneously; sustainably increasing agricultural production needs to be one of them.

The proverb "necessity is the mother of invention" proved true in the long history of agricultural development. In spite of concerns and predictions to the contrary, farmers, breeders, and other scientists always developed and implemented agricultural technologies that ensured food production growth was sufficient to meet the rising demand. From a mere technological perspective, there is no reason to believe that this would not be possible in the future as well. As will be shown in the following chapters, modern biotechnology and genetic engineering—combined with other promising technologies—could play a crucial role in increasing agricultural productivity and product quality, while reducing the use of scarce resources and chemical inputs. But nowadays the rates and directions of innovation in agriculture crucially depend on public attitudes. New technologies need to be used responsibly, but given the challenges ahead we cannot afford to rule out promising areas of science simply based on prejudices.

CHAPTER 3

POTENTIALS AND RISKS OF GM CROPS

A genetically modified (GM) crop is a plant used for agricultural purposes into which genes coding for desirable traits have been inserted through genetic engineering. The term genetic modification is somewhat misleading, as it implies that plants had not been genetically modified before techniques of genetic engineering were developed. As described in chapter 2, humans have modified the genetic makeup of plants since the beginnings of agriculture. Without the initial cultivation of plants, our cereals would still be seed-shedding wild grasses and our potatoes small, toxic lumps. Without systematic selection, our maize would be unrecognizable, and sugarbeets would not exist in their known form. Without scientific plant breeding, our crops would be relatively inefficient nutrient converters and susceptible to countless diseases and pests. All of these developments represent genetic modifications of crops, which would not have occurred naturally without human intervention. And without these interventions by breeders, agricultural yields would only be a fraction of what they are today. It is thus not the genetic modification of plants that is new, but some of the methods involved in achieving this modification (Kloppenburg, 2004; Becker, 2011).

Tools of Genetic Engineering

The term genetic engineering does not refer to the transfer of genes only, but it also includes all techniques related to identifying and characterizing genes. Genetic engineering builds on the discoveries of the genetic code in the second half of the twentieth century, especially the discovery of the double-helix structure of DNA through James Watson and Francis Crick in 1953. Techniques of identifying and characterizing genes are relatively uncontroversial, as they do not permanently alter a plant's genotype. Such techniques are well established in modern plant breeding.

In contrast, the permanent alteration of a plant's genotype by means of direct gene transfer has sparked considerable controversy since the first GM crops were developed in the 1980s and first commercialized in the mid-1990s. Interestingly, the same type of gene transfer was used in other sectors long before it finally found use in plant breeding (Murphy, 2007a). Human insulin, for instance, traditionally produced from animal insulin, was first developed using genetically engineered bacteria in 1978, a notable achievement in the treatment of diabetes. In food processing, the use of genetically engineered microorganisms has been common for many years. A significant part of the cheese consumed worldwide, is now produced using chymosin from GM bacteria, GM fungi, or GM yeasts. Chymosin is an enzyme used for the splitting of milk proteins and was previously extracted from the stomach of young ruminants. That GMOs are relatively undisputed in these other sectors but not in agriculture is an interesting phenomenon. One reason is probably that GM crops are released to the environment, whereas GM microorganisms in medicine and food processing are mostly used under contained conditions.

As discussed in chapter 2, plant breeders depend on genetic variation for the development of new, useful crop varieties. To increase the genetic variability in a species of interest, breeders have—for a long time—used wide crosses, mutagenesis, protoplast fusion, and similar techniques, which can lead to fairly random outcomes. Genetic engineering opens new horizons, as the genetic pool to choose from, and hence the variation available for breeding, becomes much larger. Individual genes that were identified to code for desirable traits can be introduced to the plant in a targeted way, without simultaneously making the many other genetic changes that occur through conventional crossing or induced mutations. Thus, the precision of developing varieties with desirable traits has increased significantly (Kempken and Jung, 2010).

When the first GM crops were developed in the 1980s, a widely held perception was that plant genetic engineering is an entirely new science, competing with conventional breeding approaches in the development of superior plant varieties. A far more accurate view, as we now know, is that genetic engineering is a new method employed by plant breeders to create genetic variability. It can therefore complement and enhance the plant breeding process but will not replace conventional tools of crop improvement (Kloppenburg, 2004).

To transfer genes from one organism to another using recombinant DNA techniques, two different mechanisms are available, namely biolistics, involving the injection of cells with foreign DNA through a gene gun, and *Agrobacterium*-mediated transformation, where the soil bacterium *Agrobacterium tumefaciens* is used as a vector of the foreign DNA.

Usually, the gene of interest is transferred together with a selectable marker to facilitate detection of whether or not the transfer was successful and the new trait is actually expressed in the recipient plant. After successful transformation, cell and tissue culture techniques are used for plant regeneration from the transformed cells. Out of large numbers of transformation events that are usually produced, the best performing are chosen for further technology development and testing.

GM crops are often referred to as transgenic crops, implying that genes from other species—so-called transgenes—were introduced. While such gene transfer is possible across species boundaries, genes from the same species can also be introduced using biolistics or *Agrobacterium*-mediated transfer. If genes from the same species are introduced with genetic engineering, the resulting plants are called cisgenic instead of transgenic (Kempken and Jung, 2010). Cisgenesis can be useful to edit certain traits without an extensive change to the genome and/or when working with species that are very difficult to breed because of not producing true seed. A case in point is potato where cisgenesis has been used to introduce late blight resistance. More generally, cisgenesis seems to be a powerful tool to furnish crops with lost properties that their ancestors once possessed in order to withstand unfavorable environmental conditions, a process that is referred to as "reverse breeding" (Palmgren et al., 2015). An interesting question that is still under debate is whether cisgenic crops should be regulated as strictly as transgenic GM crops or more in line with conventionally bred crops (Schubert and Williams, 2006; Palmgren et al., 2015). Cisgenic crops could potentially be more acceptable to the public, as it seems to be the gene transfer across species boundaries that causes particular concern.

Another recent development in genetic engineering involves new transformation techniques. With biolistics and *Agrobacterium*, genes are transferred such that the exact genome location is not predictable. This can lead to traits not being properly expressed or to other undesirable effects in some of the transformation events. Such inferior events have to be eliminated through a careful selection process. Lately, new techniques were developed that allow specific areas of plant DNA to be modified, thus further increasing the precision of the transformation process. These new techniques include zinc finger nucleases (ZFNs), transcription activator-like effector nucleases (TALENs), and clustered regulatory interspaced short palindromic repeat (CRISPR)/Cas-based RNA-guided DNA endonucleases (Townsend et al., 2009; Gaj et al., 2013; Ricroch and Hénard-Damave, 2015). Sometimes these new tools are collectively referred to as genome editing or targeted gene modification (TagMo) technologies. TagMo technologies have not yet been commercialized, and it is unclear whether they will fall under the existing GM crop regulations (Kuzma and Kokotovich, 2011). These trends show that plant genetic engineering is a very dynamic field of research. Tomorrow's techniques may be different and even more sophisticated than what is currently conceivable (Cressey, 2013).

Breeding Objectives

In chapter 2, I have described that farmers and plant breeders have tried (and managed) for millennia to develop crops with desirable traits from an agricultural production perspective. The breeding objectives now pursued with tools of genetic engineering are not different from those pursued through conventional breeding. However, with the much larger genetic variation that can be exploited, some traits that were previously difficult or impossible to breed can now be developed with relative ease. These traits can then be introgressed into locally adapted crop varieties with other desirable characteristics. The local adaptation process itself depends on long-term breeding with conventional methods, which underlines that genetic engineering is a complement, not a substitute for conventional breeding.

What are the major traits that plant breeders work on with tool of genetic engineering? In an attempt to give a broad overview, it is often differentiated between first-, second-, and third-generation GM crops, depending on the complexity of the traits and the expectation of when these technologies might become available in the market (Moschini, 2008; Qaim, 2009). Classifications are not always uniform, however, and expected timelines keep on shifting because of research and regulatory obstacles in some cases. Hence, an overview by types of traits seems more useful. Broadly speaking, one can differentiate between input traits and output traits. Input traits can generate a direct advantage in agricultural production through higher yields or lower costs, whereas output traits relate to improvements in product composition and quality.

Input Traits

Input traits that breeders work on with recombinant DNA techniques include resistance to biotic stress factors, such as pests and diseases, tolerance to abiotic stress factors, such as drought and heat, and related plant characteristics that help to increase and stabilize crop yields or reduce chemical inputs and production costs. Among the first GM traits that were developed and commercialized were herbicide tolerance (HT) and insect resistance. Genes coding for HT make the crop tolerant to certain herbicides, which can facilitate weed management in farmers' fields. The most widely used GM technology so far is Monsanto's Roundup Ready technology, which makes the crop tolerant to the broad-spectrum herbicide glyphosate. The technology's name comes from the fact that Monsanto sells glyphosate under the brand name Roundup. A few other HT technologies with tolerance to other herbicides are available as well. Insect resistance is so far primarily based on genes from the soil bacterium *Bacillus thuringiensis* (Bt), which is why these crops are often referred to as Bt crops. Different Bt genes code for resistance to different types of insects from the lepidopteran and coleopteran orders. HT and Bt crops are already widely used (see chapter 4 for further details). Other GM traits involve resistance to fungi, viruses, and bacteria. Some of these other GM resistance mechanisms are at advanced stages of development and testing in various crops (see chapter 5). GM virus resistance has already been commercialized in a few horticultural crops (James, 2014).

In many parts of the world, farmers use large amounts of chemical pesticides to control insect pests, diseases, and weeds. This pesticide use is not only expensive, but can also be harmful for the health of farmers, consumers, and the environment. Higher levels of genetic resistance in host plants, regardless of whether this is achieved through genetic engineering or other forms of breeding, can help to reduce the reliance on chemical pesticides. Hence, the introduction and wider use of pest- and disease-resistant crops could contribute to reductions in chemical pesticide use with possible economic, environmental, and health benefits. Moreover, pest- and disease-resistant GM crops may increase effective yields (Qaim and Zilberman, 2003). Pest- and disease-resistance mechanisms do not change the yield potential of crops, but they help to reduce crop losses, which can be sizeable. In spite of chemical pesticide use, crop losses occur in many situations, either because effective chemicals are not available or farmers do not have the knowledge and resources to use them effectively. It is estimated that yield losses in major field crops due to insects, diseases, and weeds amount to 30–40 percent (Oerke, 2006). Actual losses are higher in developing countries than in developed countries because pest and disease pressure in tropical and subtropical climates is often stronger than in temperate zones. And, given more severe technical and financial constraints, pest control is often less effective in developing countries.

Research on abiotic stress factors includes crop tolerance to drought, heat, flooding, and coldness, among others. Some of these traits are genetically more complex, and early attempts to introduce such tolerance mechanisms to crop plants partly led to unintended influences on other traits through so-called pleiotropic effects (Kempken and Jung, 2010).

However, recent approaches have been more successful in producing stress tolerance in plants without unintended side-effects (Baulcombe et al., 2013). The first drought-tolerant GM maize was recently commercialized in the USA (James, 2014). Other technologies are also at advanced stages of field testing. Such traits that make plants more tolerant to unfavorable weather conditions could help to stabilize yields. Developing and growing crops that can better cope with weather stress is an important adaptation strategy to climate change, which may not only cause higher mean temperatures but also more frequent weather extremes.

Researchers are also working on developing GM crops with enhanced nitrogen and phosphate use efficiency through various mechanisms (Baulcombe et al., 2013). Such technologies will allow reductions in mineral fertilizer applications without jeopardizing yields. In situations where low quantities of fertilizer are used and soils are nutrient-depleted, as in large parts of Africa, enhanced nutrient uptake by crop plants may also contribute to higher yields. A much more complex trait is the plant's ability to fix atmospheric nitrogen. Legumes naturally have this ability, which researchers also try to get into cereals. Unlike GM crops with enhanced nitrogen use efficiency, which are at an advanced stage of development, engineering of nitrogen fixation in cereals is a rather long-term but possibly high-reward objective (ENSA, 2015). Other breeding objectives related to soil conditions include plant tolerance to soil salinity and aluminum toxicity with promising initial results (Schroeder et al., 2013). Soil salinity is a particular problem on irrigated lands and can reduce crop yields significantly.

Biotechnologists are also trying to increase the yield potential of crops. While this has always been a major objective of plant breeders, genetic engineering has helped to better understand and influence the underlying physiological mechanisms in the plant (Long et al., 2015). One approach in this direction is to enhance photosynthetic efficiency. Photosynthesis is the process through which plants convert carbon dioxide to biomass (sugars) using sunlight. The efficiency of this process is hampered by photorespiration, where the plant uses oxygen instead of carbon dioxide thus reducing biomass production and yield significantly. Genes to reduce photorespiration have been identified and are tested in various plants. Several laboratories are also seeking to bring the efficient C4 photosynthetic pathway from plants like maize, to wheat, rice, and other cereals that contain the less efficient C3 photosynthetic pathway (Baulcombe et al., 2013). Engineering the C4 pathway into C3 plants is a complex undertaking and may still take many years to be successful. But the potential to increase yield without the need for more external inputs is substantial (IRRI, 2015).

Genetic engineering can also be useful to develop traits in plants that help to improve the efficiency of the breeding process itself. A case in point is male sterility, implying that the plant is incapable to produce functional pollen. Male sterility as such does not increase crop yield, but it is a very useful trait for the development of hybrid varieties that have higher yields due to hybrid vigor. As explained in chapter 2, hybrids are obtained from crossing two highly inbred parent lines, which requires that self-pollination is avoided. Male sterility can help to avoid self-pollination at low cost. While genetic male sterility is available in several crop species, development of this trait with conventional tools has proven difficult in other species, which is one reason why hybrids are not widely used in crops such as wheat. With genetic engineering breeders cannot only introduce male sterility for the production of hybrid seeds but also ensure that pollen fertility is restored in the progeny, if desired (Kempken and Jung, 2010).

Output Traits

The development of output traits includes food crops with improved quality characteristics as well as plants that produce substances that can be useful for nutrition, health, or industrial purposes. The first GM crop that was commercialized in the USA in 1994 was the so-called Flavr Savr tomato that included a gene to delay the ripening process, while allowing the tomato to retain its natural color and flavor. The Flavr Savr tomato was taken off the market in 1997 for various reasons, including low consumer acceptance of GM foods. Yet the idea to develop crops with longer shelf lives and reduced tendency for bruising is further pursued by biotechnologists. Several genes that help reduce oxidative processes have been identified and transferred. Such traits could be particularly useful to reduce post-harvest losses and waste in fruits, vegetables, and other perishable crops.

Techniques to increase the production of desired substances in plants, reduce the production of undesired substances, or introduce pathways for the production of new substances are referred to as plant metabolic engineering. Metabolic engineering offers large potentials for the food and agricultural sector and beyond. For instance, using tools of genetic engineering, researchers have managed to eliminate proteins with allergenic potential from wheat and other foods, without altering other characteristics of the plants (Baulcombe et al., 2013). Work is also ongoing to produce foods that can help to reduce the risk of type II diabetes and high blood cholesterol, or foods that contain elevated levels of health-promoting compounds such as flavonoids, fructans, and vitamins. The term "functional foods" is sometimes used to describe foods with novel characteristics to increase human health. This term is not confined to GM crops, but plant metabolic engineering offers new potentials to develop functional foods. While several GM crop technologies with improved health characteristics are at different stages of development, very few of these technologies were so far tested under field conditions.

Metabolic engineering also involves the development of plants with higher amounts of micronutrients—like minerals and vitamins—in the edible parts, which is also referred to as biofortification. Biofortification can involve both conventional breeding and genetic engineering. A well-known example that was not possible to develop without recombinant DNA techniques is Golden Rice that contains high amounts of beta-carotene, a precursor of vitamin A (Potrykus, 2001). Beta-carotene has also been introduced in sorghum and other crops. Furthermore, several projects work on introducing iron, zinc, folic acid, and other micronutrients into food crops using transgenic and conventional techniques (Qaim et al., 2007; De Steur et al., 2015). Work on elevating the amounts of several micronutrients in the same crop, which is sometimes referred to as multi-biofortification, is also fairly advanced (De Steur et al., 2012). Micronutrient deficiencies are a widespread problem, contributing to child mortality and other severe health problems, especially among poor population segments in developing countries. Biofortified crops have the potential to reduce these nutrition and health problems (Qaim et al., 2007).

Genetic engineering is also used to develop plants with traits of interest far beyond the food sector. Plant starch, for instance, is a raw material widely used for industrial purposes. Researchers can modify the metabolism in plants such that the starch produced has improved structural properties for technical uses. Work is also ongoing to produce other types of biopolymers in plants for use in fiber, fuel, and cosmetics industries (Kempken and Jung, 2010). Plants and algae are the only organisms on earth that can convert solar energy and carbon dioxide into organic compounds. Such plant technologies could reduce the need for fossil fuels and other non-renewable resources in the medium and long run.

Other uses of metabolically engineered plants are in the pharmaceutical industry. Several GM plants that produce monoclonal antibodies, hormones, and other biopharmaceuticals were already developed, although none of these technologies has so far been approved for commercial use. Many of these proteins for the pharmaceutical industry are currently produced by GM microorganisms. But global demand for diverse biopharmaceuticals increases, so that new, cost-effective production capacities may be required. From a regulatory perspective, such types of GM plants may be difficult to handle. It needs to be ensured that the pharmaceu-ticals produced in the plants do not enter the regular food chain with a zero tolerance threshold. Therefore, it is likely that plant species that are not used for food and feed purposes will eventually be chosen for product development, or that approvals for such types of GM crops will only be given for use under contained conditions (Qaim, 2009).

GM Crop Traits and Varieties

One fundamental difference between conventional plant breeding and plant genetic engineering is the product of the research. The product of conventional breeding is a new crop variety that has certain desirable characteristics and can be used by farmers in the particular environment for which it was developed. Some improved varieties are well adapted to a wide range of agroecological conditions, while others were bred for very specific soil and climate conditions. In contrast, the product of plant genetic engineering is not one particular new variety, but a GM trait. This difference has several important implications. First, GM technolo-gies will not replace conventional breeding, rather both approaches are highly complementary. Locally adapted varieties contain a large bundle of various characteristics that cannot easily be designed through genetic engineering. However, genetic engineering can be used to add individ-ual traits of interest to such locally adapted varieties. Second, GM tech-nologies can help to conserve varietal diversity (Krishna et al., 2015). Previously, when a superior new variety was developed, farmers often adopted this new variety, abandoning a larger number of old varieties and landraces. New GM traits can be introgressed into many existing variet-ies (Zilberman et al., 2007).

A third implication of the differences between conventional breed-ing and genetic engineering is that there is a clearer separation between trait development and local breeding. Genes that were identified to code for a desirable trait, such as drought tolerance or insect resistance, can be transferred to multiple varieties around the world. For instance, the same Bt genes that were introduced to US cotton varieties to control bollworms were also introduced to local cotton varieties in China, India, Pakistan, and South Africa. Hence, the basic research that is carried out by plant biotechnologists can have wide international applicability. With some adjustments in the gene constructs and transformation protocols the same genes can also be transferred to other species. For instance, the same Bt genes that are used in cotton are also used in maize and veg-etables to control for relevant insect pests. A fourth implication is that recombinant DNA techniques allow the introduction of desirable traits also to vegetatively propagated crops that are difficult or impossible to improve through conventional crossbreeding. A case in point is banana. Edible bananas are seedless clones, making conventional breeding very difficult. Genetic engineering could help to introduce disease resistance into popular banana varieties, something that has not yet been successful with conventional breeding techniques. Another important food crop that is notoriously difficult to breed is cassava.

While the first efforts of plant biotechnologists were targeted at developing GM crops with single improved traits, there are now also several GM crops with stacked traits available. Examples of stacked GM technologies include crops with several Bt genes to control for a wider spectrum of insect pests, or crops with genes coding for tolerance to different herbicides such as glyphosate, glufosinate, dicamba, and 2,4-D. SmartStax is a GM maize technology with eight stacked genes, namely six Bt and two herbicide-tolerant genes (ISAAA, 2013). As crop plants are exposed to various biotic and abiotic stresses, and quality characteristics increasingly matter in the market, the importance of gene stacking will likely grow in the future. Stacked GM crops can be developed by crossing plants that contain single transgenes, by re-transforming a plant that already contains another transgene, or by developing a gene construct with multiple genes that is introduced to the plant in a single transformation event. All of these approaches have their pros and cons in terms of research costs and regulation.

Risks of GM Crops

The development of GM crops with new traits and characteristics is associated with risks that need to be carefully assessed and managed. This is of particular importance because GM crops are living organisms that are released to the environment and consumed by humans and animals. Broadly speaking, two different types of risks can be distinguished. First, possible risks related to the plant transformation process itself. For instance, one could suspect that the direct transfer of genes might involve changes in the plant that are quite different from those induced by other breeding methods. Such risks related to the process of plant transformation itself would apply to all GM crops, irrespective of the particular genes introduced. Second, possible risks related to the particular genes and traits introduced. For instance, when a gene coding for the production of a substance that is toxic to humans is introduced to a food plant, negative health consequences can occur. Such risks related to a particular trait would equally occur if the same trait were achieved in the plant through other breeding methods.

Thirty years of risk research related to GM crops suggest that there are no significant risks related to the plant transformation process itself. In other words, GM crops are not *per se* more risky than conventional plant breeding technologies (Arber, 2010; European Commission, 2010b; EASAC, 2013; House of Commons, 2015). The same conclusion has also been drawn in statements by the World Health Organization, the American Medical Association, and various science academies, including in France and Germany, where the public opposition against GM crops is particularly strong. Based on the scientific evidence available there is no justification to regulate GM crops differently than conventionally bred crops. In reality, however, there is a huge difference in regulation. The approval of a new conventional crop variety only requires demonstration that the variety performs at least as well as existing varieties. For the approval of a new GM crop, many years of molecular, biochemical, and environmental testing are required. Some precaution when dealing with new technologies may always be advisable. But GM crops are not so new anymore; they have been widely used and consumed for 20 years without a single case of harm to human health or unexpected environmental effects. GM crops are the most highly regulated and tested foods in the world (DeFrancesco, 2013). Many crop varieties that are commonly used in conventional and organic agriculture would not have been approved if the same standards that are now used for GM crops had applied.

The second type of risk, namely those related to the particular genes and traits introduced, is different. Such risks exist, but they cannot be assessed for GM crops in general because herbicide tolerance will likely have different effects than insect resistance, drought tolerance, or traits involving changing starch structures or higher vitamin levels in foods. These specific risks can only be assessed case by case. And it is important to note that the same risks would also be present for any conventionally produced crops with the same traits. Hence, a product-based approach to risk regulation, as observed in the USA and Canada, makes more sense than the process-based approach for the regulation of GM crops in the EU (see chapter 6 for further details on GM crop regulation). We will review product-based health and environmental risks in the following sections.

Health Risks

The main health concerns related to novel foods, regardless of whether they are GM or non-GM, involve the possible toxicity or allergenicity of substances produced in the plant. If a gene that codes for the production of a substance that is toxic to humans is introduced to the plant, toxic reactions have to be expected. Therefore, genes coding for the production of known toxins or allergens should not be used for the development of GM food crops. But not all substances with toxic or allergenic potential are known or can be predicted. Hence, for every GM food crop laboratory tests and feeding studies with animals (usually rodents) are routinely carried out to test for possible physiological reactions. Such tests are required before GM crops are approved for use in agricultural practice and for human consumption. If the test results indicate any issues, the product would not be approved by the regulatory authorities. Food-safety regulations usually require animal feeding studies over a period of 90 days. While this is considered relatively short by some, several long-term studies that have been conducted confirm the safety record of GM crops (Snell et al., 2012).

A few cases where allergens were found to be expressed in GM crops were widely publicized. In one case, a methionine-rich protein from the Brazil nut was introduced into soybean with the intention to improve the nutritional value of soy meal used as animal feed. When tested it was found that this protein could cause skin reactions in humans that are allergic to Brazil nut. In another case, a gene was transferred from beans to peas where it was found to cause allergic reactions due to protein modification. These products were never approved for commercial use and not further pursued by its developers. It is important to mention that the same types of protein modifications in plants can also occur through breeding with conventional techniques. The likelihood of unintended changes in the plant is somewhat increased when transferring genes from other species, but we discussed in chapter 2 that wide crosses between varieties of different species are also possible without genetic engineering. The potential for unintended changes is even higher for mutation breeding (mutagenesis) than it is for recombinant DNA techniques (DeFrancesco, 2013). This is no reason to be particularly concerned about mutant varieties, which are widely used and not specifically regulated. Yet, this comparison shows how little public perceptions and regulatory approaches for GM crops are congruent with the scientific evidence.

A different health concern with GM crops relates to antibiotic resistance genes. Antibiotic resistance genes are sometimes used as molecular markers to detect whether the traits of interest are actually expressed in the GM crop. These marker genes serve no purpose in the final GM crop, but removing them retroactively is difficult. Antibiotic resistance genes do not pose any harm for human health as such. But there is a very low likelihood that they could be transferred to bacteria in the gastro-intestinal tract through horizontal gene transfer. In such case, resistant bacteria could possibly spread, making antibiotic treatments less effective.

Yet, even under highly artificial conditions, the likelihood of a horizontal transfer of antibiotic resistance genes is less than one in one-trillion (DFG, 2010). In newly developed GM crops either completely different marker genes or antibiotic resistances without relevance in human medicine are used, so this risk can be managed. Problems of antibiotic resistance have recently increased in human medicine, but these are completely unrelated to GM crops.

Metabolic engineering with the production of new substances in the plant for nutrition, health, and industrial purposes, can be associated with health risks that need to be carefully tested. But again, these risks are related to the product and the concrete substances produced in the plant, not the process of breeding with recombinant DNA technology. As mentioned earlier, plants that produce pharmaceuticals will need specific regulation to assess risks and ensure that these products do not enter the regular food chain.

Beyond possible negative health consequences, GM crops may also be associated with positive health effects such as reduced occupational health hazards for farmers through lower chemical pesticide use or lower pesticide residues in foods that consumers eat (Qaim, 2009). Furthermore, more effective pest control through GM technology can contribute to significantly reduced food contamination with mycotoxins, as was shown for insect-resistant Bt maize (Folcher et al., 2010). Finally, the consumption of GM foods with enhanced nutritional traits could cause considerable nutrition and health benefits, especially for poor people in developing countries who often have less access to alternatives than richer population segments (Qaim, 2010). Such effects will be discussed in more detail in chapters 4 and 5.

Environmental Risks

One of the early environmental concerns with respect to GM crops was that these crops could "escape" from the cultivated fields, multiply, and disturb natural ecosystems. This is not a realistic scenario, however, because domesticated crops—whether GM or non-GM—cannot thrive in natural environments. Millennia of selection and breeding made cultivated crop varieties well adapted to farming conditions but unable to compete in natural environments. However, there are other mechanisms how genes from domesticated species can spread to the environment, namely through vertical and horizontal gene transfer. Such spreading of genes is not specific to GM crops, but occurs in the same way also from conventionally bred varieties. Conventional varieties also carry genes that would not occur in a particular ecosystem without human intervention. The spread of genes is a common phenomenon and an integral part of evolutionary processes. This cannot be considered a risk as such. A risk can occur, however, when the foreign genes confer a particular fitness advantage to organisms in the natural ecosystem. In that case, these organisms may reproduce more successfully, potentially suppressing other organisms and species and contributing to biodiversity loss.

Vertical gene transfer is the flow of genes between organisms of the same or related species through outcrossing (pollen flow). Gene flow between domesticated crops and their wild relatives has been a constant feature of agriculture ever since people began to cultivate plants. Such vertical gene transfer requires that there are wild relatives growing in the vicinity of the cultivated fields. Over the millennia, domesticated crops were often imported by farmers and breeders from other world regions, so that native relatives are not found everywhere. For instance, neither in the USA nor in Europe wild relatives of major field crops such as maize, wheat, or soybean are found. For other examples, wild relatives exist so that outcrossing occurs. Whether this is associated with a significant fitness advantage depends on the trait in question and can only be assessed case by case. Flow of HT traits does not pose a problem for natural biodiversity because herbicides are not used in natural ecosystems. Nutrition traits, such as higher vitamin contents, are also not associated with biodiversity risks because vitamins are not required by plants to grow well.

Somewhat different is the situation for pest- and disease-resistance traits. Wild plants are also attacked by insects, viruses, fungi, and other pests, so that higher levels of pest and disease resistance obtained through outcrossing can constitute a certain fitness advantage. However, compared to intensively cultivated fields, pest and disease infestation levels tend to be much lower in natural ecosystems due to higher species diversity. Hence, wild relatives of domesticated crops that carry Bt or other pest-resistance genes are unlikely to pose a significant environmental threat (Raven, 2010).

Genes that confer tolerance to abiotic stresses may be of more concern from a biodiversity perspective, when wild relatives of the domesticated crops occur in the region. For instance, drought tolerance may constitute a significant fitness advantage for wild plants growing in dry environments. Related risks for biodiversity occur regardless of whether the new tolerance traits are from GM or non-GM crops. Such risks need to be properly assessed and managed. This can mean that particular varieties would not be approved in certain agroecological conditions.

Clearly, vertical gene transfer also occurs between GM and non-GM domesticated crops of the same species. For instance, pollen from GM maize can pollinate non-GM maize growing on the neighboring field.

This is not an environmental risk; when the GM maize itself was found to be safe, the cross-pollinated maize will also be safe. However, such outcrossing can become an economic risk for organic farmers because the harvest with the transgenes cannot be sold as organic anymore, so the organic price premium is lost. Existing organic standards explicitly exclude GMOs from certified organic production. Such economic risks can be reduced through coexistence rules, such as minimum distances between fields grown with GM and non-GM crops of the same species (Beckmann et al., 2006).

Gene flow can also occur through horizontal gene transfer, referring to the exchange of genes across species boundaries by means other than sexual reproduction. Sometimes, plants incorporate foreign DNA from microorganisms through natural horizontal gene transfer (Kyndt et al., 2015). More often, horizontal gene transfer occurs the other way around, that is, bacteria and viruses incorporate foreign genes from plants and other organisms. Except for antibiotic resistance that was already described earlier, traits that are useful in crop plants are very unlikely to confer a fitness advantage to bacteria, viruses, and other microorganisms. Hence, while horizontal gene transfer can occur in both GM and non-GM crops, environmental risks are very low and were not yet found to be of any practical relevance (DFG, 2010).

Beyond gene flow, GM crops can possibly have unintended effects for non-target organisms. Bt crops, for instance, produce Bt proteins that are toxic to lepidopteran and coleopteran insects. The Bt proteins are harmless for other insects, and also for higher-order animals and humans, which is why Bt-based biological pesticides are also widely used in organic agriculture. Studies that were carried out under artificial laboratory conditions showed that a few non-target insect species might also be affected by Bt, including monarch butterflies and ladybirds. Follow-up research showed that such effects for non-target insects are unlikely to occur under field conditions (Romeis et al., 2008, 2014). In any case, Bt is much less harmful for non-target organisms than most chemical pesticides. Hence, the prevalence of beneficial insects in fields cultivated with Bt crops was found to be significantly higher than in fields cultivated with conventional crops (Wolfenbarger et al., 2008). Also for organisms in the soil and in aquatic environments, no negative effects of Bt were found under field conditions (Bartsch et al., 2010). For non-GM varieties, effects on non-target organisms have never been analyzed so extensively.

Possible resistance buildup in pest populations is another concern regarding GM crops. Resistance development is a common problem in pest control, also without GM crops. Given natural genetic variation, a few individuals in a pest population are always likely to have genetic resistance against a particular pesticide or a pest-control mechanism inbuilt into crop varieties. Selective pressure can lead to an increase in the frequency of resistance and thus to the pesticide or other pest-control strategy to become less effective. There are many pesticides, such as synthetic pyrethroids, that have gradually lost their effectiveness over time. The same can also happen for GM crops. For Bt crops, resistance buildup was observed in certain target pests and regions after several years of using this technology (Tabashnik et al., 2013). However, the observed increase in the frequency of Bt resistance is locally confined, occurred slower than initially expected, and did not yet lead to a significant increase in chemical pesticide use, suggesting that Bt technology has not yet lost its effectiveness.

There are several reasons why the problem of Bt resistance buildup is still largely under control. First, in most countries where Bt crops were commercialized farmers have to grow a certain proportion of their area with non-Bt crops of the same species, to provide a refuge for pest populations and thus reduce selective pressure (Romeis et al., 2008). Second, especially in diverse smallholder systems, where several crops are often grown on small fields side by side, unintended refuge areas often exist. Many Bt target pests have several host plants that they feed on, which can also reduce selective pressure when some of these host plants are non-Bt. For instance, cotton bollworms also feed on maize, sorghum, soybeans, several legumes, and a wide variety of vegetables. This may be one reason why Bt resistance buildup did not happen more widely in developing countries where formal refuge regulations do not exist or are difficult to monitor (Krishna and Qaim, 2012). Third, most Bt crops nowadays have two or more stacked Bt genes that produce different toxins. Gene stacks involving different pest-control mechanisms make rapid resistance buildup less likely (Tabashnik et al., 2013).

This positive experience with resistance to Bt crops—or lack thereof in most cases—should not be used to downplay the general problem of possible resistance buildup. The issue needs to be tested and managed also for other GM and non-GM crops with inbuilt mechanisms to control pests and diseases. However, the possibility of resistance buildup is sometimes used as a knockout argument against GM crops. This is inconsistent because the same argument could also be used against any other pest- and disease-control strategies in agriculture and human medicine.

Resistance to herbicides in weeds is a particular concern regarding HT crops. The issue is somewhat different than for resistance buildup in Bt target pests. For HT crops, weeds do not develop resistance to the crop technology itself, but to the herbicide that the crop can tolerate. In many places of North and South America, weed species have developed resistance to glyphosate because of the widespread use of this herbicide year after year in glyphosate-tolerant soybean and maize crops (Fernandez-Cornejo et al., 2014). Selection pressure could be reduced if farmers would also use other herbicides from time to time, but they prefer glyphosate for convenience. As mentioned earlier, HT has more recently also been developed for herbicides other than glyphosate—including glufosinate, 2,4-D, and dicamba—as individual and stacked technologies. Use of these other technologies with different modes of action may help to slow down further resistance buildup to glyphosate. Glyphosate is considered relatively benign because it is less prone to leaching than other herbicides, more biodegradable, and less toxic to animals and humans (Service, 2007).

As another environmental problem, opponents to GM crops also mention that the technology may facilitate or contribute to monocultures, with the same crop species grown season after season. Monocultures are usually bad for the environment as they tend to decrease soil quality and increase pest and disease problems. The trend toward monocultures in some regions is driven by farmers' profit considerations and began long before the first GM crops were commercialized (Barrows et al., 2014). Ensuring proper crop rotations, adjusted to local conditions, is important from a sustainability perspectives but can hardly be achieved through banning GM crops. This will require rules of good agricultural practice and a wider recognition of the fact that breeding technologies cannot substitute for good agronomy.

In summary, all forms of agriculture affect natural ecosystems and biodiversity. The available evidence clearly shows that the ecological problems related to the cultivation of GM crops fail to differ in any fundamental way from those associated with conventional crops or agricultural production in general, except that GM crops often involve the application of lower amounts of chemicals (Raven, 2010).

When comparing the environmental effects of different forms of agriculture, the reference chosen is of crucial importance. One hectare of extensive agriculture with low amounts of external inputs has fewer disruptive effects for the ecosystem than one hectare of intensive agriculture. However, one hectare of extensive agriculture also produces significantly lower output, so that more land is required to satisfy the given demand for food and other agricultural produce. Cropland is limited, and converting more pristine land into agricultural production comes with high environmental costs in terms of biodiversity loss and greenhouse gas emissions. Hence, in a world with rising global demand, the environmental costs of different forms of agriculture cannot be compared only per hectare. There are important environmental tradeoffs between extensive and intensive forms of agriculture, and the nature of these tradeoffs changes with changing framework conditions and technological progress. GM technology offers potentials to reduce such tradeoffs through improved crops that produce higher yields while reducing the need for external inputs and thus the environmental impact of intensive production systems.

Social Risks

Apart from health and environmental risks, GM crops may also be associated with social risks, such as widening disparities between rich and poor. Social risks are different from health and environmental risks, as they are usually not inherent to a technology but rather relate to questions such as: who develops the technology? Who is able to use it successfully? And how is the technology regulated? Leisinger (1995) has used the term "technology-transcending risks" for issues that are not inherent to a technology but emanate from its mode of application in certain circumstances. Technology-transcending risks are best dealt with by altering the external conditions for the better, for instance, through improved policies and institutions.

Technology-transcending social risks include concerns that high-tech applications such as GM crops may benefit developed countries more than developing countries, large farms more than small farms, and rich consumers more than poor ones. There are also fears that patents on crop technologies may contribute to unfair seed prices, new dependencies, industry concentration, and corporate control of the entire food chain. These concerns are only mentioned here because evidence on economic and social impacts is discussed in much more detail in the chapters to follow.

CHAPTER 4

ADOPTION AND IMPACTS OF GM CROPS

While in chapter 3, potentials and risks of GM crops were discussed, this chapter takes a closer look at what we know about actual impacts in different situations. As GM crops have already been used commercially for 20 years, a large number of impact studies exist, looking at GM crop effects on farmers' yields, pesticide use, income, poverty, and wider implications for sustainable development. I will first provide an overview of the adoption of GM crops in different parts of the world, before reviewing the evidence about impacts. Impacts can be analyzed *ex post*, based on actual observations, or *ex ante*, based on expert assumptions and simulations of likely future scenarios. In this chapter, I review *ex post* impact studies of already commercialized GM crops. A few *ex ante* studies for future GM crop applications are discussed in chapter 5.

Global Adoption of GM Crops

The commercial application of GM crops began in the mid-1990s. Since then, the technology has spread rapidly around the world, both in developed and developing countries (Figure 4.1). Since 2011, the area grown with GM crops in developing countries has been larger than that in developed countries. In 2014, GM crops were planted on 182 million hectares, which is equivalent to 13 percent of the total worldwide cropland. These 182 million hectares were grown by 18 million farmers in 28 countries (James, 2014). The countries with the biggest shares of the total GM crop area were the United States (40%), Brazil (23%), Argentina (13%), India (6%), Canada (6%), China (2%), and Paraguay (2%). Among the countries of the European Union (EU) only Spain grows GM crops at significant scale. In other EU countries, the area grown with GM crops is negligible, mainly because of public acceptance problems and unfavorable regulatory frameworks.

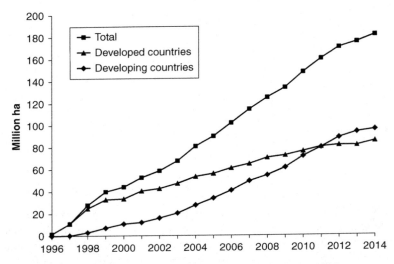

Figure 4.1   Worldwide area cultivated with GM crops (1996–2014).

Source: Own presentation with data from James (2013; 2014).

In spite of the widespread international use of GM crops, the portfolio of available crop-trait combinations is still very limited. While many of the traits discussed in chapter 3 were already developed and tested, most of them were not yet approved for commercial use because of lengthy regulatory procedures and extremely cautious approaches by policymakers who are responsive to public prejudices against GMOs. So far, only a few concrete GM technologies have been commercialized. The dominant technology is herbicide tolerance (HT) in soybeans, which made up 50 percent of the global GM crop area in 2014. HT soybeans are currently mostly grown in the United States, Brazil, Argentina, several other Latin American countries, and South Africa. This technology now accounts for 82 percent of total worldwide soybean production.

GM maize is the second-most dominant crop and covered 30 percent of the worldwide GM area and 30 percent of total maize production in 2014 (James, 2014). GM maize involves HT and insect resistance, partly as separate and partly also as stacked technologies. Insect resistance is based on different genes from the soil bacterium *Bacillus thuringiensis* (Bt). These Bt genes in maize control the European corn borer, the corn rootworm, and different stemborers (Romeis et al., 2008). Bt maize is mostly grown in North and South America, but is also planted to a significant extent in South Africa and the Philippines.

Other GM crops with significant area shares include cotton, canola, and sugarbeet. Bt cotton with resistance to bollworms is particularly relevant in developing countries. In 2014, India had the largest Bt cotton area with 11.6 million hectares, followed by the United States and China with around 4 million hectares each, and Pakistan with 3 million hectares. Several other developing countries also grow Bt cotton, including Argentina, Brazil, Burkina Faso, Mexico, Myanmar, South Africa, and Sudan. In the United States and Argentina, Bt and HT cotton are employed, partly with stacked genes. HT canola and HT sugarbeet are mostly grown in Canada and the United States up till now.

There are also a few other GM crops that have been approved in individual countries, so far only covering relatively small areas. These other GM crops include HT alfalfa, as well as virus-resistant papaya, squash, sweet pepper, and tomato.

Global Overview of GM Crop Impacts

A large number of studies have analyzed the impacts of commercialized GM crops on chemical pesticide use, yield, and profits of adopting farmers using micro-level data collected in various countries (Qaim, 2009; Smale et al., 2009; Carpenter, 2010; Finger et al., 2011; Areal et al., 2013). A recent meta-analysis of the original studies was recently conducted, to consolidate the available evidence (Klümper and Qaim, 2014). As most impact studies focused on GM soybean, maize, and cotton with HT and insect resistance (IR) traits (based on Bt genes), the meta-analysis also focused on these crop-trait combinations. A total of 147 original studies were identified through keyword searches in relevant scientific literature databanks. All these studies compared the performance of GM crops with the performance of non-GM crops of the same species and in the same environment. While reported impacts differ by concrete GM technology and country, the meta-analysis revealed clear overall trends that are summarized here. More specific details about HT and IR crops are discussed in subsequent sections.

On average, GM technology has increased crop yields by 21 percent (Table 4.1). These yield increases are not due to higher genetic yield potential, but to more effective pest control and thus lower crop damage. At the same time, GM crops have reduced chemical pesticide quantity by 37 percent and pesticide cost by 39 percent. Pesticides here refer to all chemical pest control agents. Of particular relevance for HT and IR crops are herbicides and insecticides. The effect of GM crop adoption on the cost of production per hectare is not significant. GM seeds are usually more expensive than non-GM seeds because GM seeds are predominantly

Table 4.1 Mean impacts of GM crop adoption (in %)

Outcome variable	Total sample	Insect resistance (IR)	Herbicide tolerance (HT)
Yield	21.6***	24.9***	9.3**
Pesticide quantity	−36.9***	−41.7***	2.4
Pesticide cost	−39.2***	−43.4***	−25.3***
Total production cost	3.3	5.2**	−6.8
Farmer profit	68.2***	68.8***	64.3

Note: Results are based on a meta-analysis of 147 original studies analyzing the impact of GM soybean, maize, and cotton in different parts of the world. **, *** statistically significant at 5% and 1% level, respectively.

Source: Own presentation with data from Klümper and Qaim (2014).

sold by private companies at a premium. But the additional seed costs are compensated through savings in chemical and mechanical pest control. Average profit gains for GM-adopting farmers are 68 percent. These are very sizeable positive effects for farmers and the environment.

Table 4.1 also shows a breakdown by modified crop trait. While significant reductions in pesticide costs are observed for both HT and IR crops, only IR crops cause a consistent reduction in pesticide quantity. Such disparities are expected because the two technologies are quite different. IR crops protect themselves against certain insect pests, so that spraying can be reduced. HT crops, on the other hand, are not protected against pests but against a broad-spectrum chemical herbicide (mostly glyphosate), use of which facilitates weed control. While HT crops have reduced herbicide quantity in some situations, they have contributed to increases in the use of broad-spectrum herbicides elsewhere (see below for further details). The savings in pesticide costs for HT crops in spite of higher quantities can be explained by the fact that broad-spectrum herbicides are often much cheaper than the selective herbicides that were used before. The average farmer profit effect for HT crops is large and positive, but not statistically significant because of considerable variation.

Klümper and Qaim (2014) also estimated meta-regression models to analyze other factors that might influence the reported impacts of GM crops. Consistently, the gains in yield and farmer profit and the savings in pesticide costs through GM crop adoption are higher in developing countries than in developed countries. Reasons for these higher benefits in developing countries are twofold. First, especially smallholder farmers in the tropics and subtropics suffer from considerable pest damage that can be reduced through GM crop adoption (Qaim and Zilberman, 2003). Pest infestation levels tend to be lower in temperate zones (Oerke, 2006). Moreover, farmers in developed countries are better educated and equipped with alternative pest control technologies. Second, most GM crops are not patented in developing countries, so that GM seed prices are lower and own seed reproduction by farmers is possible. There are widespread public concerns that GM crops may be more suitable for rich countries and that farmers in developing countries may not benefit or may be driven into new dependencies. Yet the evidence suggests that farmers in developing countries benefit even more than their colleagues in developed countries.

Most original impact studies included in the meta-analysis build on farm surveys of randomly selected adopters and non-adopters of GM crops (Klümper and Qaim, 2014). A few of the original studies were also based on field-trial data. Field-trial results are often criticized to overestimate impacts because farmers may not be able to replicate experimental conditions. While this is true in general, significant differences between GM crop impacts reported from surveys and from field trials could not be observed. Reported yield gains from field trials are even lower than those from farm surveys. This is plausible, as pest damage in non-GM crops is often more severe in farmers' fields than on well-managed experimental plots. In the meta-regressions it was also analyzed whether the statistical method of data analysis plays a role for reported GM crop impacts. Many of the early impact studies simply compared yields and other performance indicators of GM and non-GM crops without controlling for differences in other inputs and conditions. This can lead to what Stone (2012) has called "cultivation bias," because there may be differences in the use of fertilizer, irrigation, and other factors that also affect observed outcomes. Several more recent impact studies controlled for other relevant factors through regression approaches. Interestingly, results derived from regression approaches lead to higher average yield effects of GM crops, suggesting a downward bias in earlier studies that did not control for other factors. For GM crop impacts on pesticide use and farmer profits, the statistical method of data analysis seems to matter less (Klümper and Qaim, 2014).

A concern often voiced in the public debate is that studies funded by industry money might report inflated benefits. Companies that develop and sell GM seeds have a vested interest in positive study outcomes. When the same companies finance impact analyses, it is possible that they somehow influence the researchers involved implicitly or explicitly. Of the 147 original studies included in the meta-analysis, over 90 percent were funded by the public sector, where no funding bias is expected. Moreover, the meta-regressions revealed that the source of funding does not significantly influence the direction or magnitude of the impact estimates.

Finally, it was analyzed whether the type of publication matters. Many of the 147 studies included in the meta-analysis were published in peer-reviewed academic journals, others were published as conference papers, book chapters, or institutional reports. Interestingly, studies published in peer-reviewed journals show significantly higher yield and profit gains than studies published elsewhere. On first sight, one might suspect publication bias, meaning that only studies that report substantial effects are accepted for publication in a journal. However, further analysis suggests that the journal review process does not systematically filter out studies with small effects. The journal articles in the sample report a wide range of yield effects, including negative estimates in some cases. It is rather likely that the papers published elsewhere may suffer from a downward bias. Indeed, studies that were not published in a journal encompass a diverse set of papers, including reports by NGOs and outspoken biotechnology critics. These reports show smaller GM crop benefits on average, but not all meet common scientific standards in terms of sample size and representativeness. Instead of sampling randomly for data collection, NGOs that are opposed to GMOs sometimes collect data only in locations or from farmers that they know suffered from crop failures due to drought or other unfavorable conditions (Stone, 2012). Such studies can create a lot of negative publicity for GM crops, but they would not usually survive a rigorous peer-review process by a good academic journal.

A meta-analysis—such as discussed here—is useful to get a consolidated picture of GM crop impacts from a bird's eye view, but not necessarily to understand the micro-level effects of concrete GM crop applications in specific contexts. I will therefore present additional micro-level evidence in the following sections, differentiating between the impacts of HT and IR crops.

Impacts of Herbicide-Tolerant Crops

HT crops are tolerant to certain broad-spectrum herbicides. Most HT crops grown so far are tolerant to the herbicide glyphosate, which Monsanto sells under the brand name Roundup. Glyphosate-tolerant crops developed by Monsanto are therefore also called Roundup Ready crops. Glyphosate controls almost all types of weeds, even at higher growth stages. Glyphosate also affects crop plants, so that in conventional agriculture this particular herbicide can only be applied before the crop emerges. With HT crops, glyphosate can be applied at any time. Glyphosate is more effective than the selective herbicides that are used in conventional farming. Glyphosate is also less toxic than most selective herbicides. It has very little residual activity and is rapidly decomposed to organic components by microorganisms in the soil. According to the international classification of pesticides by the World Health Organization (WHO), it belongs to toxicity class IV, the lowest class for "practically non-toxic" pesticides. Finally, glyphosate is cheaper than most other herbicides. Monsanto's patent on glyphosate already expired in the 1990s, so that there are now also several other companies selling generics in the United States and elsewhere. And indeed, as the meta-analysis discussed earlier already indicated, farmers who adopted HT crops and switched from selective herbicides to glyphosate to control weeds, benefit from lower herbicide costs (Qaim, 2009).

However, given the price differences, lower herbicide costs do not necessarily imply lower herbicide quantities. After the introduction of HT crops, the use of glyphosate has increased rapidly, while the use of other herbicides has declined. Total herbicide quantities applied were reduced in some situations, but not in others. In Argentina, herbicide quantities were even increased significantly (Qaim and Traxler, 2005). This is largely due to the fact that herbicide sprays were substituted for tillage. In Argentina, the share of soybean farmers using no-till soil conservation practices doubled to almost 90 percent since the introduction of HT technology. Also in the United States and Canada, no-till practices were expanded significantly with HT crop adoption (Fernandez-Cornejo et al., 2014; Smyth et al., 2014).

Yield effects of HT crops are also diverse. In some situations, there is no significant yield difference between HT and conventional crops, which suggests that farmers were already controlling weeds quite effectively before adopting GM seeds. In other situations, where weed infestation is higher or certain weeds occur that are difficult to control with selective herbicides, the adoption of HT and the switch to broad-spectrum herbicides resulted in better weed control and higher crop yields. Examples for significant yield advantages are HT soybeans in Mexico and Bolivia, HT maize in Argentina and the Philippines, and HT canola and sugarbeet in the United States and Canada (Brookes and Barfoot, 2013).

Overall, HT technology reduces the cost of production through lower expenditures for herbicides, labor, machinery, and fuel. Yet, since HT crops were developed and commercialized by private companies, a technology fee is charged on seeds. This fee varies between crops and countries. Several early studies for HT soybeans in the United States showed that the fee was in a similar magnitude or sometimes higher than the average cost reduction, so that farmer profit effects were small or partly negative. Comparable results were also obtained for HT cotton and HT canola in the United States and Canada (Qaim, 2009; Smyth et al., 2014). The main reason for farmers in such situations to still use

HT technologies was easier weed control and the saving of management time. Fernandez-Cornejo et al. (2005) showed that the saved management time for US soybean farmers partly translated into higher off-farm incomes. Moreover, farmers are heterogeneous, that is, many adopters have benefited in spite of zero or negative mean gross margin effects. The average farm level profits seem to have increased over time, partly due to seed price adjustments, farmer learning effects, and better HT varieties becoming available.

In South American countries, the average farmer profit effects of HT crops are larger than in North America. While the agronomic advantages are similar, the fee charged on seeds is lower, as HT technology is not patented there. Many soybean farmers in South America even use farm-saved GM seeds. Average profit gains through HT soybean and maize adoption in Argentina, Brazil, Colombia, Paraguay, and Uruguay are in a magnitude of 20–80 US$ per hectare (Qaim and Traxler, 2005; Brookes and Barefoot, 2013). HT is so attractive for farmers that adoption rates in countries where this technology was commercialized are very high. In Argentina and Paraguay, close to 100 percent of the national soybean areas are grown with HT varieties. Also in the United States, where the monetary benefits for farmers are somewhat lower on average, adoption rates of HT soybean, maize, and sugarbeet are above 90 percent.

Beyond the financial impacts for farmers, HT crop adoption also has some broader implications. The lower toxicity of glyphosate compared to other herbicides was already mentioned. The much wider use of no-till practices that was facilitated by HT technology adoption has also produced environmental benefits in terms of lower fuel use, lower soil erosion, lower greenhouse gas emissions, and higher soil quality. On the other hand, there are also some environmental problems associated with HT crop use. One problem is related to the fact that HT crops are now sometimes cultivated as monocultures, without any crop rotation. Monocultures can reduce soil quality and increase pest and disease problems over time. It is not recommended to grow HT crops as monocultures, but given the higher profitability and convenience of this technology farmers do so nonetheless. A related problem is the development of glyphosate resistance in various weed species in North and South America. Resistance development is due to the fact that most farmers who adopted HT crops used glyphosate applications as the only weed control strategy season after season. This has caused constant selection pressure on weed populations and thus the evolution of glyphosate resistance.

Farmers in regions where significant glyphosate resistance occurs have increased their glyphosate use and have started again to include other herbicides, which are often more toxic. Some farmers have also resumed tilling their soil as an additional weed control strategy. Thus, some of the initial environmental benefits of HT crops are lost. GM crops that are tolerant to other broad-spectrum herbicides and stacked tolerance traits can help to address the problem of resistance. Several new HT technologies with tolerance to other herbicides were recently commercialized (James, 2014). But it is also important to learn from past mistakes. Sustainable technology use requires better crop and technology management by farmers, with suitable crop rotations and alternating weed control strategies before resistance problems occur. This may require policies that define and enforce rules of good agricultural practice.

The high adoption rates of HT crops in various countries may also raise concerns about the potential loss of varietal diversity. Because the Roundup Ready technology with resistance to glyphosate was developed by Monsanto, it is sometimes falsely assumed that only a few Monsanto varieties were available, replacing a larger number of non-GM varieties by other seed companies. However, as explained in chapter 3, once developed GM traits can be introgressed into a large number of varieties through backcrossing. And this has actually happened. In the United States, several hundred HT soybean, maize, and canola varieties are used, well adapted to local conditions and sold by various seed providers that all licensed HT technology from Monsanto and other biotech companies (Qaim, 2005; Barrows et al., 2014). Also in Brazil and Argentina, several hundred HT varieties are used. As Monsanto's Roundup Ready technology is not patented there, seed providers can backcross this trait into their varieties also without a formal license.

In the public debate about GM crops the question of what type of farmers can successfully use the new technologies is often raised. Especially in developing countries, it is of interest whether smallholder farmers can benefit because the small farm sector is home to a large proportion of the poor. This depends on several factors, the suitability of the particular technology for smallholder conditions, the affordability of GM seeds, and the wider infrastructure and institutional context, such as access to roads, technical information, credit, and output markets. For HT technology, the experience in the small farm sector is limited. While farmers in South America and South Africa benefit significantly from HT soybeans, most soybean growers are relatively large and fully mechanized farms. Soybean is not a typical smallholder crop, and other HT crops are not yet widely used in developing countries. Smallholders often weed manually because labor is cheap. In such situations, adoption of HT crops would not make sense because this technology is only useful in combination with particular chemical herbicides. Switching to chemical herbicides would reduce the demand for manual labor and could contribute to rural underemployment.

However, in some developing countries, cheap rural labor is getting scarce, so that a switch to chemical weed control might be practical. Also, the adoption of natural resource management technologies, which could help to increase the sustainability of smallholder systems, could be facilitated by a switch to chemical weed control. A case in point is conservation agriculture, which is more difficult to implement without the use of herbicides. Another example is the system of rice intensification (SRI), which can save water and reduce greenhouse gas emissions from flooded rice fields. SRI involves growing rice on moist but not flooded fields, a shift that tends to increase problems with weeds (Noltze et al., 2013). Finally, in some situations there are weeds that are very difficult to control manually. For instance, *Striga* (witchweed) causes significant crop losses in maize and sorghum among smallholders in Africa and Asia. This discussion suggests that HT technology may be suitable for some, but not for all smallholder situations.

Impacts of Insect-Resistant Crops

Insect-resistant Bt crops have different effects than HT crops. Bt crops produce proteins that are toxic to larvae of some lepidopteran and coleopteran insect species. Therefore, Bt is a pest control agent that can be used as a substitute for chemical insecticides. If pest pressure is high and farmers use a lot of chemical insecticides in the conventional crop, Bt adoption will lead to substantial insecticide reductions. However, Bt technology can also have significant impacts on effective crop yields. While Bt genes do not affect yield potential of plants, they can lead to a reduction in crop losses, when there is previously uncontrolled insect pest damage. Insecticide reduction and yield effects are closely related: farmers who use little amounts of insecticides in their conventional crop in spite of high pest pressure will realize a sizeable yield effect through Bt adoption, while the insecticide reduction effect will dominate in situations where farmers initially use higher amounts of chemical inputs.

The meta-analysis of GM crop impacts discussed above confirmed that both insecticide-reducing and yield-increasing effects can be observed internationally for Bt technologies. We now look at the situation in different countries more specifically. The upper part of Table 4.2 shows average effects of Bt cotton. Reductions in chemical insecticide use range from 21 percent in Pakistan to 77 percent in Mexico. As expected, yield effects are more pronounced in developing countries, due to higher pest infestation levels in the tropics and subtropics. The highest yield effects

Table 4.2   Average effects of Bt crops on insecticide use, yield, and farmer profit

Country	Change in insecticide use (%)	Change in crop yield (%)	Change in farmer profit (US$/ha)
	Bt cotton		
Argentina	−47	33	23
Australia	−48	0	66
Burkina Faso	−67	20	64
China	−65	24	470
India	−50	24	192
Mexico	−77	9	295
Pakistan	−21	28	504
South Africa	−33	22	91
USA	−36	10	58
	Bt maize		
Argentina	0	10	20
Philippines	−5	34	53
South Africa	−10	11	42
Spain	−63	6	70
USA	−8	5	12

Source: Own presentation with data from Qaim (2009), Krishna and Qaim (2012), Kathage and Qaim (2012), Kouser and Qaim (2013), Vitale et al. (2014).

are observed in Argentina, where conventional cotton farmers under-use chemical insecticides, so that insect pests are not effectively controlled (Qaim and Janvry, 2005). The lower part of Table 4.2 shows results for Bt maize. The patterns are similar, albeit the effects are lower in magnitude than for Bt cotton. In general, insect pest problems are more severe in cotton than they are in maize.

While the country level average effects shown in Table 4.2 clearly underline the agronomic and economic advantages of Bt crops, they mask the fact that impact variability is also observed within countries and over time. The suitability of insect-resistant Bt crops depends on local pest infestation levels, which can vary regionally and seasonally. In China, for instance, infestation levels of cotton bollworms are highest in the northern and eastern parts of the country, so that the benefits of Bt cotton are most pronounced there. This is reflected in much higher adoption rates, as compared to Western China. In the United States, due to diverging pest infestation levels, Bt crop adoption rates are lower in California than in other states (Qaim, 2009). Impact variability of Bt cotton in India is discussed further below.

Beyond Bt maize and cotton, preliminary evidence on the impacts of other Bt crops also exists from field trial results and limited commercial applications. In China, Bt rice to control stemborer and leaf roller has been developed and tested for several years in farmer-managed preproduction trials (Huang et al., 2005). The Chinese authorities granted a biosafety certificate for Bt rice in 2009, with approval for full commercialization still pending. One of the reasons why Bt rice was not yet fully commercialized in China are concerns about losing certain export markets. The EU would probably not approve GM rice for imports.

Based on the preproduction trial data, Wang et al. (2012) showed that the yield gains of Bt rice in China could be up to 65 percent in situations where no chemical insecticides are used. Yet, most farmers in China use insecticides quite intensively, so that actual yield gains under commercial conditions will be lower. Instead, significant insecticide reductions are possible. Under realistic farming conditions in China, insecticide reductions of Bt rice are in a magnitude of 50–60 percent (Huang et al., 2005; Wang et al., 2012).

Bt eggplant to control the shoot and fruit borer has been developed and tested for several years in India, Bangladesh, and the Philippines. Combining two-year field trial data with farm survey data from India, Krishna and Qaim (2008a) showed that Bt eggplant could reduce insecticide use by 40 percent, while raising yields by 40–60 percent under typical farming conditions. The Indian biosafety authorities had approved Bt eggplant in 2009, but the influence of biotech critics contributed to the government's decision to suspend the use of this technology for an indefinite period of time (Herring, 2015). Meanwhile, Bangladesh has commercialized Bt eggplant in 2013. In 2014, vegetable farmers in Bangladesh started using this technology. Together with cotton, vegetables are the types of crops that are most heavily treated with chemical pesticides in Asia and elsewhere.

Unlike herbicide-tolerant crops that have not yet been widely adopted in the small farm sector and may also not be suitable for all types of smallholder situations, insect-resistant Bt crops are already extensively used by smallholder farmers. Bt cotton in particular is used by millions of small farms in India, China, Pakistan, Burkina Faso, and other developing countries. In South Africa, many smallholders grow Bt white maize as their staple food. Several studies show that Bt technology advantages for small-scale farmers are of a similar magnitude as those of larger-scale producers. In some cases, the advantages can be even greater because smallholders often have less access to effective pest control alternatives due to knowledge and financial constraints (Qaim, 2009). Smallholders may also not always have proper access to Bt technology, depending on the type of crops involved and seed market conditions (Qaim and de Janvry, 2003). However, the evidence available so far shows that Bt crops can be a very promising option for smallholder farmers in developing countries. Ali and Abdulai (2010) showed that Bt cotton adoption has contributed to poverty reduction in Pakistan. Subramanian and Qaim (2010) and Kathage and Qaim (2012) found similar effects in India; the case of Bt cotton in India will be discussed in more detail in a separate section further below.

The significant reductions in chemical insecticides through Bt crop adoption are also associated with environmental and health benefits. Insecticide reductions are particularly sizeable in cotton. Brookes and Barfoot (2014) estimated that—between 1996 and 2012—Bt cotton was responsible for a global saving of 205 million kg of pesticide active ingredients, reducing the environmental impact of total cotton pesticides by 28 percent. Especially in developing countries, Bt adoption leads to over-proportional reductions in the most toxic insecticides (Qaim, 2009; Abedullah et al., 2015). Lower insecticide use means lower soil and water pollution and higher biodiversity in the agricultural system. It also means less insecticide exposure for farmers and farm workers during spraying operations. This is especially relevant for smallholder farmers in developing countries, who mostly apply insecticides manually with little information about the negative health effects.

Studies showed that Bt cotton adoption among smallholders significantly reduced the frequency of acute farmer pesticide poisoning in China, India, Pakistan, and South Africa (Bennett et al., 2003; Hossain et al., 2004; Kouser and Qaim, 2011, 2013). Significant positive health effects for farmers are also expected for Bt rice and Bt eggplant, once these technologies are approved and widely used (Krishna and Qaim, 2008a; Huang et al., 2015). Kouser and Qaim (2013) carried out a choice experiment with cotton farmers in Pakistan, in order to value the environmental and health benefits of Bt technology in cotton production. They estimated an aggregate benefit of 195 US$ per hectare, of which 53 percent was due to environmental improvements and 47 percent to health improvements. These environmental and health benefits are in addition to the direct financial benefits for Bt adopting farmers.

Bt technology can also have important health benefits for consumers. One mechanism is through lower pesticide residues in food. Cotton is not eaten, but Bt eggplant and other Bt vegetables have significant potential in this respect. In India, farmers spray eggplant up to 30 times per season, causing significant problems with pesticide residues in food (Krishna and Qaim, 2008b). Another mechanism of consumer health benefits is through lower mycotoxin contamination of food. In a variety of field studies, Bt maize was shown to contain significantly lower levels of certain mycotoxins, which can cause cancer, birth defects, and other diseases in humans (Wu, 2006; Folcher et al., 2010). Especially in maize, insect damage is one factor that contributes significantly to mycotoxin contamination. In the United States and other developed countries, maize is carefully inspected, so lower mycotoxin levels might primarily reduce the costs of testing and grading. But in many developing countries, strict mycotoxin inspections are uncommon. In such situations, Bt technology can contribute to lowering the actual health burden (Parrott, 2010).

In the first years of Bt crop deployment, it was predicted that insect populations would soon develop Bt resistance, which would undermine the technology's effectiveness and lead to declining insecticide reductions over time. Rapid resistance development is possible with constant selection pressure, as was shown in several laboratory studies. However, as discussed in chapter 3, rapid resistance development did not occur under field conditions, which is largely due to successful resistance management strategies, such as the planting of non-Bt refuges (Tabashnik et al., 2008). In some cases, resistance buildup was observed. However, so far such resistance has been very localized and confined to individual species within the broader spectrum of lepidopteran and coleopteran insects (Tabashnik et al., 2013). Cotton, for instance, is attacked by various species all belonging to the broader category of bollworms, such as cotton bollworm, American bollworm, pink bollworm, spotted bollworm, tobacco budworm, and also several armyworm species. These are all Bt target pests. When one of these species, say pink bollworm, has developed resistance in one location, it does not mean that other bollworm species, or other locations, would also be affected. Most Bt varieties of cotton and maize are now equipped with two or more stacked Bt genes. Hence, when resistance to one Bt gene has developed, the other Bt genes would still provide some pest protection for the plant. A significant increase in insecticide use against bollworms, which one would expect with widespread resistance development, has not yet been observed.

The absence of widespread Bt resistance development so far does not imply that it could not happen in the future. As for any pest control strategy, proper resistance management will be necessary to provide insect protection on a sustained basis. In the past, Bt refuge requirements were mostly implemented such that farmers had to plant a certain fraction of their area with non-Bt varieties. This, however, requires an active decision by each farmer and strict monitoring. As Bt varieties are more profitable than non-Bt varieties, farmers have an incentive to not follow the refuge rules and plant the entire area with Bt. Especially in the small farm sector of developing countries, monitoring compliance is difficult and not all developing countries actually have established refuge area require-ments. While selection pressure for Bt resistance is lower in smallholder systems, due to the cultivation of other host plants for the pest species on nearby fields, this may possibly change when additional Bt crops are developed and released.

One solution is to switch from refuge area requirements to a system of "refuge in the bag" (RIB), meaning that Bt seeds are mixed with a certain percentage of non-Bt seeds already by the seed provider. Hence, farmers who decide to purchase and use Bt seeds would automatically implement the refuge requirement. First experiences with the RIB sys-tem in the United States and India are promising.

Beyond resistance, there are other factors that can lead to changes in observed Bt impacts over time. In China, for instance, insecticide applications against non-Bt target pests increased after several years of Bt cotton use (Wang et al., 2008). Such secondary pests include mirids, mealybugs, and other sucking insects. Their frequency in Bt fields has somewhat increased because of the significant reduction in chemical pes-ticide sprays. Bt farmers reduced their sprays against bollworms, but most chemical insecticides are less selective than Bt, so that the same chemicals do also control sucking pests to some extent. Using several years of field trial data from China, Lu et al. (2010) confirmed that secondary pest populations had increased in Bt cotton. However, building on nationally representative panel data covering the period 1997–2012, Qiao (2015) showed that the economic benefits of Bt cotton in China in terms of insecticide savings and farmer profit gains continue.

Krishna and Qaim (2012) analyzed pesticide use patterns in India over a period of seven years. They found that farmers who adopted Bt cotton have somewhat increased their insecticide sprays against sucking pests. Nevertheless, total insecticide use in Bt cotton further decreased over time because the rise in sprays against secondary pests was more than offset by the continued decline in sprays against bollworms. Krishna and Qaim (2012) found that conventional cotton growers in India could reduce their sprays as well because the widespread adoption of Bt led to area-wide suppression of bollworm populations. Similar effects were reported for Bt cotton in China and Bt maize in the United States (Wu et al., 2008; Hutchison et al., 2010). Such area-wide suppression is comparable to the effect of vaccination in human medicine and can be interpreted as a positive externality of Bt crop adoption for conventional farmers.

It is sometimes argued that Bt technology may not be the only option to reduce chemical insecticide use in crop production. In some regions, insecticides are overused in conventional farming, entailing a disruption of beneficial insects and increasing pest levels (Pemsl et al., 2008; Kouser and Qaim, 2014). In such cases, insecticide reductions would be possible without any loss in productivity. More careful pest scouting and biological control measures—such as promoted in integrated pest management (IPM) programs—could also help to cut down chemical insecticide use. However, IPM is labor and knowledge intensive, so that it is not yet widely adopted in smallholder agriculture. In any case, IPM and Bt technology are highly complementary approaches (Romeis et al., 2008), so that pursuing one should not be seen as a substitute for the other.

The Case of Bt Cotton in India

I will now focus on the experience with Bt cotton in India, adding additional details that are of relevance for a comprehensive assessment of technology impacts. India is an interesting case to study for various reasons. India is the country with the largest Bt cotton area worldwide, and cotton in India is mostly grown by smallholder farmers with less than 5 hectares of land. In fact, many of the cotton-growing families in India have less than 1 hectare of cotton and suffer from problems of poverty and malnutrition. Bt cotton in India is also a case that has aroused a lot of controversy, both locally and internationally (Glover, 2010; Herring, 2010; Stone, 2011). While some argue that Bt cotton is a success story, others claim that GM seeds create new dependencies and drive farmers into suicide (Shiva et al., 2011; Coalition for a GM-Free India 2012). In this section, I will review the scientific evidence about Bt cotton impacts in India. Wider politics and persistent narratives in the public debate will be discussed in chapter 7.

Bt cotton was commercially approved in India for the first time in 2002. The original Bt technology was developed by Monsanto, and Monsanto had collaborated with the Indian seed company Mahyco to adjust it to Indian conditions and incorporate the trait into local cotton hybrids. India is one of the few countries where cotton hybrids are grown; in most other countries cotton is grown in the form of open-pollinating varieties (OPVs). In 2002, around 35,000 hectares were planted with Bt hybrids. In subsequent, years the area under Bt increased substantially and reached 11.6 million hectares in 2014—around 95 percent of the total Indian cotton area. Around eight million smallholder farmers in India have adopted Bt cotton by now (James, 2014). The widespread adoption of Bt technology in India has contributed to a significant boost in the country's cotton sector. Between 2002 and 2014, India more than doubled its share in global cotton production from 12 percent to 25 percent. India has recently overtaken China to be the largest cotton producer worldwide.

In 2002, the first year of official Bt cotton cultivation in India, only three Bt hybrids, which were developed by Mahyco with Monsanto's Bollgard I technology (containing the Bt Cry1Ac gene), were approved by the national regulatory authorities. The same three Bt hybrids were also available in 2003. In 2004, a fourth Bt hybrid developed by Rasi Seeds, another Indian company that had sublicensed the Bollgard I technology, was approved. In 2005, two additional seed companies received approval for the commercialization of Bt cotton hybrids. In 2006, the number of approved Bt hybrids further increased significantly. Moreover, new Bt events were deregulated by the national authorities, including Monsanto's Bollgard II technology (containing two Bt genes, namely Cry1Ac and Cry2Ab) and competing technologies developed by public research institutes. These new Bt events were also backcrossed into hybrids from several local seed companies. In 2008, India's Central Institute for Cotton Research released the first Bt OPV. By 2014, over 1,000 Bt cotton hybrids were available in India, provided by more than 40 different seed companies (James, 2014). This development underlines that nowadays Indian cotton farmers have a large choice of hybrids in a highly competitive seed market.

Together with my research team, I have carried out research on the impacts of Bt cotton in India for many years. We have collected data from several hundred randomly selected cotton farms in four states that are representative for the central and southern cotton belts of India—the regions where most of the smallholders operate. In northern India, farms are somewhat larger on average. We collected data every two years in four rounds always from the same farms, thus creating a panel database covering a period of seven years (Kathage and Qaim, 2012; Krishna and Qaim, 2012). This is a unique database for the evaluation of Bt cotton impacts. Many other impact studies build on cross-section data from only one year. Panel data have the advantage that they allow the analysis of trends over time, which is not possible with cross-section data.

Panel data also help to better deal with issues of selection bias in impact assessment. Selection bias occurs when adopters and non-adopters of a new technology are systematically different in terms various characteristics. For instance, technology adopters may be smarter and have better access to technical information, so that their yields might be higher even without the new technology. In that case, simply comparing crop yields of technology adopters and non-adopters would overestimate the technology's impacts. Controlling for observed differences between adopters and non-adopters through a regression approach is useful, but does not suffice in the presence of unobserved confounding factors. This is where the panel structure of the data helps: since data from the same farms are available before and after adoption, statistical differencing techniques can be employed to reduce selection bias and provide more accurate impact estimates.

Our results in terms of the effects on yields, insecticide use, and farmer profits are line with other scientific studies that have been carried out on Bt cotton impacts in India (Bennett et al., 2006; Crost et al., 2007; Rao and Dev, 2010). Yet, due to the breadth of the information collected in our surveys and the panel structure of the data, we could also analyze broader welfare effects and impact dynamics, which the other studies could not.

In the first season of technology use in 2002–2003, we found significant reductions in chemical insecticide use and significant gains in yields and profits for Bt-adopting farms on average (Qaim et al., 2006). However, we also found large regional differences. While farmers in three of the states surveyed (Maharashtra, Karnataka, Tamil Nadu) benefited over-proportionally, Bt-adopting farms in the fourth state (Andhra Pradesh) experienced negative yield and profit effects. The reason for the negative experience in Andhra Pradesh was not that Bt technology failed to work there, but that the technology was incorporated into hybrids that were not well adapted to the local soil and climate conditions. Hence, any benefits of the technology were overshadowed by a yield drag because of unsuitable germplasm. As mentioned, in the beginning only three Bt hybrids were available in India, which were not perfectly adapted to all locations. Even though additional Bt hybrids had been developed already at that stage, the national authorities were hesitant to approve them because they wanted to observe the performance of a small number of Bt hybrids first. To avoid additional losses, many of the initial Bt adopters in Andhra Pradesh switched back to conventional hybrids in the following year (Qaim, 2005).

This first-season experience in India has a couple of broader lessons to offer. First, like for any technology the impacts of Bt technology can vary regionally depending on local conditions. Second, to optimize the benefits it is important to incorporate GM traits into locally adapted varieties or hybrids. Third, regulatory hurdles such as the non-approval of additional GM hybrids can have detrimental effects for farmers and can also jeopardize varietal diversity. Fourth, farmers make rational decisions about adopting and dis-adopting GM crops based on their expectations and personal experience. The fact that farmers who made losses in one season switched back to conventional seeds in the following season underlines that there is no dependency that forces farmers to stick to GM seeds. With the substantial increase in the number of Bt hybrids available in India in subsequent years, farmers in all regions could benefit and varietal diversity was restored (Krishna et al., 2015). Farmers in Andhra Pradesh also re-adopted Bt technology, this time incorporated into hybrids well adapted to their conditions.

The panel data results reveal substantial benefits of Bt adoption over the entire 7-year period, with average yield gains of 24 percent and farmer profit gains of 50 percent (Kathage and Qaim, 2012). While temporal variability exists due to variation in pest pressure, a declining trend in the gains—as hypothesized by many due to possible resistance development or secondary pest outbreaks—was not observed. Some of the panel model estimates even suggest increasing benefits over time, which is actually plausible given that additional Bt events with a broader spectrum of target pest species and a larger number of Bt hybrids became available. Bt seed prices have declined since 2006 due to government price interventions in some of the Indian states and increasing competition among seed companies. The Bt effects on chemical insecticide use are shown in Figure 4.2. Bt adoption did not only lead to lower insecticide use, but insecticide use on Bt fields also further declined over time. This is in spite of the fact that farmers had to spray a bit more against secondary pests. With rising Bt adoption rates, the few remaining conventional cotton growers could also reduce their sprays due to area-wide suppression of bollworms. Kouser and Qaim (2011) showed that

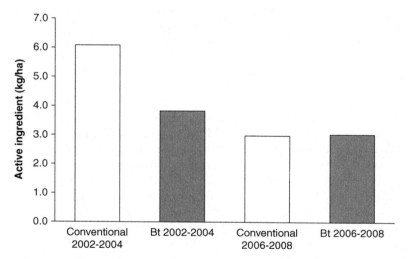

Figure 4.2 Insecticide use in India in Bt and conventional cotton (2002–2008).

Source: Own presentation with data from Krishna and Qaim (2012).

the widespread use of Bt cotton in India has reduced farmer pesticide poisoning by over two million cases per year.

Since cotton production is the main source of income for most small-holder cotton farmers in India, the profit gains through Bt adoption also contribute to rising household living standards. A common way of measuring living standard in the development economics literature is through household consumption, including the value of food consumption from own production and market purchases as well as non-food consumption of goods and services. We used this measure and showed that Bt adoption has increased household living standards by 18 percent on average (Kathage and Qaim, 2012). This has also translated to improved nutrition and dietary quality. Through the higher income from cotton sales, adopting household increased their food energy and micronutrient consumption by 5–10 percent. Bt technology has reduced food insecurity among cotton-producing households by 15–20 percent (Qaim and Kouser, 2013).

Millions of smallholder farm households in India could increase their incomes and living standards through Bt cotton adoption. Such broad-based developments are also likely to have wider effects for the local economy through consumption and production spillovers. On the consumption side, higher household incomes cause higher expenditures on goods and services thus contributing to demand-led growth. On the production side, higher cotton yields increase the demand for farm labor employed for harvesting. Higher harvests also mean more employment in transport and processing sectors. Such spillovers can be analyzed with general equilibrium models. We constructed a detailed social accounting matrix (SAM) of the village economy based on comprehensive census data from one typical cotton-growing village in the state of Maharashtra. We then developed a SAM-multiplier model to simulate the direct and indirect effects of Bt cotton adoption on the rural economy (Subramanian and Qaim, 2010). Our results show that Bt technology is indeed employment generating, especially for female agricultural laborers. Effects on rural household incomes, including farm households and landless rural laborers, are shown in Figure 4.3. Each additional hectare of Bt cotton produces 82 percent higher aggregate incomes than conventional cotton, implying remarkable economic growth in the village economy. All types of households—including those below the poverty line—benefit more from Bt than from conventional cotton. Sixty percent of the aggregate benefits accrue to the extremely and moderately poor. These findings demonstrate that Bt cotton contributes to poverty reduction and rural development in India.

How do these results fit together with the NGO claims of a causal link between Bt cotton adoption and farmer suicides in India? In fact,

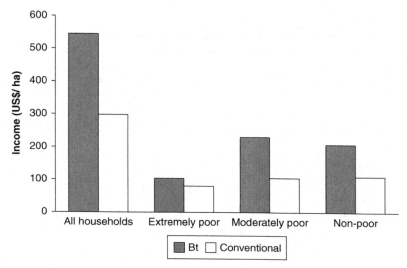

Figure 4.3   Effects of Bt cotton on rural household incomes in India.

Note: Results are based on a SAM-multiplier model for a typical village economy, taking into account direct effects of Bt cotton adoption as well as indirect effects through consumption and production spillovers.

Source: Own presentation with data from Subramanian and Qaim (2010).

they don't. The suicide claims contradict the evidence, but they are nonetheless perpetuated by the mass media and are very powerful in the public debate. To be sure, thousands of suicides are observed every year among Indian farmers, and many of them occur in the country's cotton belts. These suicides are tragic, but they seem to be caused by a variety of factors. Erratic rainfalls and related crop failures and indebtedness seem to be important components (Patel et al., 2012). But suicides among farmers in India have been reported long before Bt cotton was introduced, and the rates have not increased since 2002 (Figure 4.4). Hence, there is no direct link between suicides and Bt technology. What is also not widely known is that suicide rates in India are among the highest in the world. Suicides among farmers account for less than 10 percent of all suicides in India (Gruère and Sengupta, 2011; Patel et al., 2012), meaning that suicide rates among farmers are actually lower than in other population groups. The overall problem surely needs attention, but this is an issue that predates and extends far beyond GM crop policies. The argument in connection with Bt cotton is pure anti-biotech propaganda.

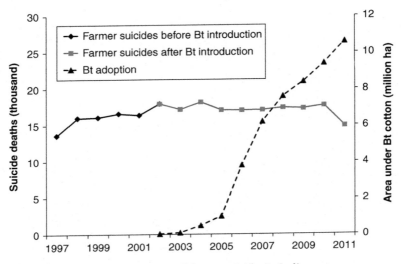

Figure 4.4   Bt cotton adoption and farmer suicides in India.

Source: Own presentation with data from Gruère and Sengupta (2011), Hesselbarth (2013), James (2014), and Qaim (2014).

Other GM Crops

While herbicide-tolerant and insect-resistant crops account for close to 100 percent of the present area cultivated with GM crops worldwide, a few other technologies were commercialized and are or were used by farmers in the United States and elsewhere. One noteworthy example is virus-resistant papaya, which was developed jointly by Cornell University and the University of Hawaii.

Papaya is an important crop in Hawaii grown by large and small farms for local use and for exports to Canada, Japan, and other countries. The papaya ringspot virus is a serious disease that can devastate papaya plantations and entire industries. After virus outbreaks on Oahu Island in the 1950s, papaya production became almost impossible there. The industry relocated to Hawaii Island. Once plants are infected with the ringspot virus, treatment is not possible. Attempts to breed virus-resistant papaya varieties with conventional methods remained unsuccessful. In the 1980s, methods were developed to produce virus-resistant GM plants by introducing the coat protein gene from the virus itself. This concept was first demonstrated for tomato and tobacco. Cornell University and the University of Hawaii used this technology to develop GM papaya varieties with resistance to the papaya ringspot virus. The virus reached

Hawaii Island in the 1990s, before GM varieties had been commercial-ized. By 1997, the main production regions on Hawaii Island were severely infected and total production had dropped to 50 percent. In 1998, two resistant GM papaya varieties, which had been field tested for several years in Hawaii, were released to farmers and rapidly adopted. Within a few years, the majority of the papaya farmers had adopted virus-resistant GM varieties. Yield levels moved back to their previous levels. So far, this technology has remained effective for more than 15 years, without any signs of resistance development in the virus strains (Gonsalves and Gonsalves, 2014).

Papaya is also grown in many other countries. Especially in Thailand, Vietnam, the Philippines, and other countries in Asia and Latin America, papaya is consumed a lot and is an important source of micronutrients. Papaya ringspot virus causes significant damage in most production regions. Cornell University has trained and supported researchers from various other countries to adapt the GM virus-resistance technology to their local conditions. Several GM varieties were tested in these countries and were found to be effective and safe. However, none of these varieties was commercially approved due to successful campaigns by anti-biotech activists (Gonsalves and Gonsalves, 2014). The only place outside the United States where virus-resistant papaya is grown is China. In China, GM papaya with resistance to local strains of the ringspot virus was devel-oped by the South China Agricultural University and commercialized in 2006 (James, 2014).

One incident that happened in Thailand with GM papaya is particu-larly noteworthy, as it demonstrates the power of European NGOs in influencing international biotechnology developments. In Thailand, 90 percent of papaya is consumed domestically; the rest is exported as canned fruit salad to various countries, including in Europe. Papaya ring-spot virus is the greatest limitation for papaya production in Thailand. In cooperation with researchers at Cornell University, the Thai Department of Agriculture developed local virus-resistant GM varieties. These variet-ies were tested in Thailand between 1999 and 2004. In 2004, Greenpeace activists destroyed one of the field trials by pulling out the plants and throwing them into biohazard bins. The activists wore personal protec-tion suits with gloves and respiratory masks; they also carried banners with the slogan "Stop GMO Field Trials" (Davidson, 2008). This inci-dent attracted a lot of media attention and set into motion a countrywide moratorium on all field testing of GM crops. The moratorium did not only affect the new papaya varieties but also all other GM crop technolo-gies that were in the R&D pipeline. Since the 1980s, Thailand had been a regional leader in developing plant biotechnology and GM crops. The

2004 incident thwarted many of the plant biotech R&D activities in the country.

Another GM technology that was commercialized but was not able to realize its full potential was insect-resistant and virus-resistant potato in the United States. The Colorado potato beetle is an important insect pest in potato production responsible for significant crop damage and high chemical insecticide applications. In 1995, Bt potato developed by Monsanto with inbuilt resistance to the Colorado potato beetle was the first Bt crop approved in the United States for commercial use. Potato farmers soon started to adopt Bt varieties under the brand name NewLeaf. Shortly after that, NewLeaf Plus varieties were commercialized with combined resistance to the Colorado potato beetle and the potato leafroll virus. Although producers were keen to use these technologies, the GM potato varieties soon disappeared from the market because of fears of consumer opposition. In 1999, McCain, the world's largest frozen potato processor, announced that it would stop to purchase GM potatoes for processing. The company's president was quoted as saying: "We think genetically modified material is very good science but, at the moment, very bad public relations" (Ryan and McHughen, 2014, p. 845). Other processing companies followed suit, so that potato farmers had no chance but to abandon GM varieties.

Macro-Level Welfare and Distribution Effects of GM Crops

Results on GM crop impacts discussed so far are based on micro-level data collected through farm surveys and field observations. But GM crops are now grown on 182 million hectares worldwide, so that impacts are also observable at the macro level. Sexton and Zilberman (2012) evaluated these macro-level effects. Based on several years of data they estimated cross-country regressions, where the production quantities of different agricultural crops in a country were explained by total land area and area grown with GM crops. In all these regressions, GM crop area had large positive effects, implying that GM technology adoption has increased country-level agricultural output. For soybean, the average production-increasing effect in GMO-adopting countries was 13 percent, for GM canola it was 25 percent, and for maize and cotton it was 46 and 65 percent, respectively (Sexton and Zilberman, 2012).

Not all of these production increases are necessarily net yield gains of GM technology. Technology-adopting farmers sometimes also change their production practices and their use of other inputs. In some cases, better weed control with GMOs has allowed farmers to grow a second crop per year, as is partly observed for HT soybeans in South America.

But GM technology triggered these effects, so the technology already contributes to considerable global production increases. Barrows et al. (2014) have estimated that 20 million hectares of additional land would have been necessary to produce the harvest of soybean and maize absent the production-increasing effects of GM technologies. Without GM cotton and canola, another five million hectares would have to be added. As land use changes account for a large share of anthropogenic greenhouse gas emissions, the land-saving effect of GM technology contributes to climate change mitigation. It also helps to avert biodiversity loss associated with agricultural land expansion into natural ecosystems.

Technology-induced production increases also contribute to effects on market prices. Without GM crops, world market prices of agricultural commodities would be significantly higher than with GM crops. For soybean and maize, average prices would be 5–10 percent higher than they actually are if HT and Bt crops had not been used. In times of scarce agricultural supplies and high food prices, as in 2008, the price-reducing effect of GM technology was even much higher (Sexton and Zilberman, 2012). Lower prices benefit consumers, as food becomes more affordable. This is especially relevant for poor consumers in developing countries, who often spend more than 70 percent of their total income on food. Lower food prices help poor consumers to improve their nutrition and overall living standards (Qaim and Kouser, 2013). Needless to say that the positive food security effects could be higher if more GM food crops were commercialized and used.

The aggregate economic effects of new technologies in a country or region, or for the world as a whole, can be evaluated with market equilibrium models (Alston et al. 1995; Moschini and Lapan, 1997). A very simple market equilibrium model is shown graphically in Figure 4.5, where D is the market demand curve for a particular crop, and S_0 is the market supply curve before the introduction of the new GM technology. In the initial situation without the technology the equilibrium price is p_0. Now the new GM technology is introduced. Adoption of productivity-increasing seeds will reduce the marginal cost of production, causing the crop's supply curve to shift downward to S_1. The new equilibrium price is p_1, which is lower than p_0. Based on this market model, the aggregate benefits for consumers and producers (farmers) of the crop can be evaluated. Economists call these effects changes in consumer and producer surplus. Consumers clearly benefit from the price decrease; the gain in consumer surplus can be calculated as area ($a+b+c$). For farmers, the price decrease leads to a loss in the magnitude of area a. However, for technology-adopting farmers this loss is lower than the gain through

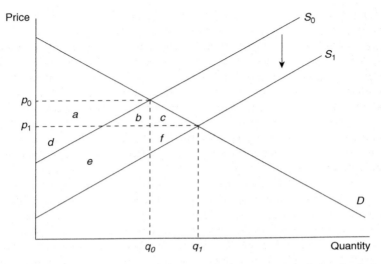

Figure 4.5   Market equilibrium model with adoption of a new GM technology.
Source: Own presentation.

the marginal cost reduction in the magnitude of area $(e+f)$. Hence, the change in producer surplus is area $(e+f)$ minus area a.

In addition to these consumer and producer surplus effects, the technology-developing company can capture an innovation rent through a technology fee charged on seed sales, especially when the technology is patented. The sum of the changes in consumer surplus, producer surplus, and company rents can be interpreted as the total welfare impact of the new technology.

When only the market of one single crop is considered, so-called partial equilibrium models are used for the evaluation. Partial equilibrium models can only capture the first-round effects of a new technology. When indirect effects and spillovers to other markets and sectors of the economy shall be captured as well, general equilibrium models are used. General equilibrium models are also often employed to capture welfare effects of international trade. Both modeling approaches have been used for the evaluation of macro-level welfare and distribution effects of GM crops. For a comprehensive review of this work, see Qaim (2009) and Frisvold and Reeves (2014).

Individual country studies showed that Bt cotton and Bt maize have each generated welfare gains of several hundred million dollars per year in the United States. The aggregate benefits have increased over time with rising technology adoption rates. Of the total gains, 30–50 percent accrued to farmers, around 20 percent to consumers through lower prices, and 30–50 percent to the innovating companies through seed price premiums. In China and India, annual welfare gains of Bt cotton were estimated above one billion dollars, respectively, with most of the benefits accruing to cotton farmers. In China, farmers captured over 90 percent of the benefits. Cotton consumers did not benefit much in the beginning because the government controlled output markets, thus preventing a price decrease. Markets were somewhat liberalized more recently, so that Chinese consumers now also benefit from Bt cotton technology. Innovation rents for companies are negligible in China, because IPR protection is weak, and use of farm-saved Bt cotton seeds widespread.

In India, Bt cotton is also not patented. However, unlike China where open-pollinated Bt cotton varieties are used, in India Bt technology is primarily used in hybrids and use of farm-saved seeds is low. Thus, private sector innovation rents are somewhat higher in India than in China. In the first years of Bt cotton adoption in India, private companies captured about one-third of the total welfare gains. This share has shrunk with more companies entering Bt cotton seed markets. Also in India, farmers are the main beneficiaries of Bt cotton technology.

There are also a number of studies available looking at the partial equilibrium effects of HT soybeans. Most of these studies use multi-region models, meaning that the effects are not only evaluated in individual countries but for the world as a whole, which is subdivided into several regions. In a recent study, Alston et al. (2014) calculated that HT soybean technology has contributed to global welfare gains of 47 billion US$ during the period from 1997 to 2010. While the benefits were relatively small in the early years with low rates of technology adoption, they are now in a magnitude of over 5 billion US$ per year. Of the total gains, around 50 percent accrues to soybean farmers in the countries where HT technology was commercialized. Farmers in other countries have no access to this technology and suffer from the reducing effect on world market prices. Consumers capture 35 percent of the total gains. Large parts of these consumer benefits accrue in countries that import soybeans, including China and the EU. Biotech companies capture about 15 percent of the gains through innovation rents. These rents only accrue in those countries where HT soybean seeds are used. Unsurprisingly, these private sector rents are significantly higher in the United States, where HT soybean technologies are patented, than in South America, where the same technologies are not patented.

These results underline a few important aspects that are not widely appreciated in the public GMO debate. First, Monsanto and other biotech companies benefit from the adoption of GM crops that they have developed and commercialized, but their share of the benefits is lower than often assumed. Many believe that private companies are the only beneficiaries of GM crop developments, which is obviously not the case. Second, farmers benefit significantly, especially in developing countries where IPR protection of breeding innovations is usually weak. Third, consumers also capture a sizeable portion of the overall welfare gains through lower prices of food and textiles.

CHAPTER 5

NEW AND FUTURE GM CROP APPLICATIONS

The previous chapter has shown that the cultivation of GM crops has increased rapidly during the last 20 years with sizeable areas in North and South America, Asia, and to a lesser extent in Africa. However, of the 182 million hectares under GM crops in 2014, 99 percent were grown with only four different crop species (soybean, maize, cotton, and canola) and two modified traits (herbicide tolerance and insect resistance). Almost all of the GM crops available so far were developed and commercialized by the private sector.

The relatively narrow focus of GM crop applications up till now has different reasons. One reason is that certain traits are more complex to develop than herbicide tolerance or insect resistance; these two traits are coded by only one single gene each. Different HT and Bt genes with different modes of action are now stacked, but the genetics are still much simpler than for other traits. Most tolerances to abiotic stress factors and many relevant quality traits are coded by multiple genes, making the process of genetic engineering much more complex. However, the more important reason for the narrow crop and trait focus so far is the low public acceptance of this technology and, coupled with this, the complex regulatory systems. Several GM technologies were extensively tested but never approved for commercial use because of overly precautious regulators, highly politicized policy processes, and extensive lobbying efforts of anti-GMO pressure groups. Examples are virus-resistant papaya in Thailand, Bt eggplant in India, or Bt rice in China (see chapter 4). Other technologies were tested and approved for commercial use, but withdrawn from the market due to concerns of low consumer acceptance. Examples are the Flavr Savr tomato (chapter 3) and insect- and virus-resistant potato (chapter 4). The German company BASF decided to withdraw its Amflora potato from the European market in 2012 due to lack of public acceptance. The Amflora potato with improved starch properties for industrial purposes was commercially approved in 2010. Other readily developed technologies were shelved by biotech companies for fear of consumer boycotts. A case in point is HT wheat that was ready to be commercialized by Monsanto in 2004.

The complex and politicized processes of biosafety and food safety regulation do not only delay the final approval and commercialization, but also the development of new GM crops, as in most countries each field trial needs a separate approval. Field trials are important to test the agronomic performance of GM crops, select preferred events for further development, and produce material required for feeding trials. When approvals for field trials are not issued on time, or when field trials are vandalized by anti-GMO activists, GM crop and trait developments can be seriously delayed or thwarted altogether. Thus, the public opposition could well contribute to a self-fulfilling prophecy: some of the public resistance is based on the argument that the promises of GM crops have been oversold because so far only very few technologies by a handful of multinational companies are actually available. Further details of regulatory processes, public concerns, and the wider implications of the anti-biotech propaganda will be discussed in chapters 6 and 7. In this chapter, I discuss some of the technologies that may make it through to farmers' fields in the next few years in spite of the many regulatory hurdles.

A few new GM technologies were recently commercialized in the United States, namely HT alfalfa and HT sugarbeet accounting for almost 1 million and 0.5 million hectares, respectively. Smaller areas of virus-resistant squash are also grown in the United States. In 2014, a new GM alfalfa technology with lower levels of lignin was commercialized in the United States. This new application helps to increase fodder quantity and digestibility of the perennial alfalfa crop. Also in 2014, Innate potato was commercially approved in the United States (James, 2014). Innate technology was developed by Simplot, a US agribusiness company, and features a quality trait to reduce the bruising of fresh potatoes. Moreover, the engineered genes help to significantly reduce the formation of acrylamide during the processing of potato. Acrylamide, a potential carcinogen, is formed naturally when potatoes are fried at high temperatures to produce French fries or crisps. While genetic engineering was used to develop this technology, no genes from species other than potato were introduced, which also explains the chosen brand name "Innate." Recent research carried out at Iowa State University showed that US consumers are willing to pay more for these GM potatoes with positive health effects than for conventional potatoes.

Another GM potato technology, for which an application for commercial approval was already filed in the United States, is late blight resistance. Potato late blight, which is caused by the fungus *Phytophtora infestans*, is the most important potato disease worldwide, causing significant yield losses. Late blight was also the cause of the Great Famine in Ireland in 1845. Since that famine, plant breeders have tried to develop late-blight-resistant potato varieties, so far without success with conventional means. Late blight is responsible for significant fungicide applications in potato production. In organic agriculture, where synthetic fungicides are prohibited, solutions of copper—a heavy metal—are often used to control potato late blight. GM late-blight resistance could reduce toxic fungicide applications and increase effective yields. This technology is also of high relevance for developing countries. Especially in some parts of Latin America and Africa, potato is an important staple food.

The first GM drought-tolerant maize technology was approved for commercial use in the United States in 2011. The drought-tolerant trait was developed by Monsanto in collaboration with BASF Plant Science and commercialized under the brand name DroughtGard. This technology is targeted at dryland maize production in the Western Great Plains of the United States, where it was already cultivated on 275,000 hectares in 2014 (James, 2014). DroughtGard does not increase yield under average or good rainfall conditions, but it is effective in reducing yield losses under moderate and severe drought conditions. DroughtGard event MON87460, which was deregulated in the United States, was also donated by Monsanto for use by African smallholder farmers under the Water Efficient Maize for Africa (WEMA) Project (see below).

Also in a few other countries, new GM crop technologies were commercialized recently or will likely be commercialized in the next one or two years. This includes technologies developed by the public sector or through public–private partnerships. The recent commercialization of Bt eggplant in Bangladesh was mentioned in chapter 4. Another example is GM virus-resistant beans, which were approved by the biosafety authorities in Brazil. This virus-resistance technology was developed by EMBRAPA, the national agricultural research organization in Brazil. Beans are an important staple food in several Latin American countries, but viruses, especially the golden mosaic virus, can cause significant production losses. To reduce losses, farmers often spray large quantities of chemical insecticides, which cannot cure virus infections but control the white fly, an important vector of viruses. The new GM technology could therefore increase effective yields and reduce chemical insecticide sprays. Farmer adoption of virus-resistant beans in Brazil is expected from 2016 onward, when sufficient seeds of suitable varieties have been produced. In Vietnam, several GM maize technologies involving HT and Bt traits were recently commercialized. In Indonesia, the first GM drought-tolerant sugarcane was recently approved by the national bio-safety authorities (James, 2014).

Emerging GM Technologies

In order to get an impression of the GM technologies that will likely be commercialized in the next 5–10 years, it is useful to take a closer look at the field trials carried out with GM crops in different countries. Such an overview is provided in Table 5.1. The list of technologies shown is not complete; in many cases, one entry involves several technologies, as is the case for insect resistance with various Bt genes or for herbicide toler-ance with multiple modes of action. Nevertheless, Table 5.1 indicates that the types of GM crops, GM traits, research organizations, and countries involved are gradually getting more diverse.

While most of the GM crop technologies commercialized so far were developed by private companies, many of the ongoing field trials are carried out by public sector organizations, often based on technologies developed in-house. Several technologies also involve public–private partnerships. Especially in Brazil and China, public sector organiza-tions have invested significantly in GM crop research during the last 10 years. But also in other developing countries, the biotech capacity and the number of GM field trials that are approved and implemented are increasing.

In terms of the crops involved, the four crops that dominated GM technology developments so far (soybean, maize, cotton, and canola) will continue to play a central role, but several other crops seem to be gaining in importance. These include a few crops that are often grown by small-holder farmers in developing countries, such as banana, cassava, cowpea, and sorghum. Such crops are often referred to as "orphan crops," because they have long been neglected by public and private research organiza-tions. The advantage of genetic engineering is that useful technologies developed in one crop can also be transferred to other crop species that suffer from similar constraints. Examples are Bt-based insect-resistant traits that were successfully transferred to rice, wheat, banana, various legumes (chickpea, cowpea, pigeonpea), and vegetables (cabbage, egg-plant, tomato) where they also provide effective resistance to lepidopteran and coleopteran insect pests. Herbicide tolerance, initially developed for soybean and maize, was also transferred to various other species, includ-ing alfalfa, rice, wheat, sugarbeet, and sugarcane.

In terms of the GM traits currently tested in the field, insect resistance and herbicide tolerance continue to play an important role. However, Table 5.1 also shows an increasing diversity of other crop traits, including

Table 5.1 Selected GM crop technologies at field-trial stage

Crop	Trait	Type of research institution	Countries
Apple	Reduced bruising/browning	Private sector	Canada
Banana	Provitamin A content	Public sector	Uganda
	Bacterial resistance	Public sector	Uganda
	Insect/nematode resistance	Public sector	Uganda
Bean	Virus resistance	Public sector	Brazil
Cabbage	Insect resistance	Public sector	China, India
Canola	Herbicide tolerance with multiple modes of action	Private sector	Australia, USA, Canada
	Omega-3 content	Private sector	USA
	Nitrogen use efficiency	Private sector	USA
Cassava	Virus resistance	Public sector	Kenya, Indonesia, Uganda
	Provitamin A content	Public sector	Nigeria, Kenya, Uganda
Chickpea	Insect resistance	Public–private partnership	India
Cotton	Stacked insect resistance and herbicide tolerance	Private sector	Burkina Faso, Cameroon, Ghana, India, Kenya, Malawi, Pakistan, USA
Cowpea	Insect resistance	Public–private partnership	Burkina Faso, Ghana, Nigeria
Eggplant	Insect resistance	Public–private partnership	India, Philippines
Groundnut	Virus/fungal resistance	Public sector	India
Maize	High phytase (quality)	Public–private partnership	China
	Stacked drought tolerance and insect resistance	Public–private partnership	Kenya, South Africa, Uganda
	Stacked insect resistance and herbicide tolerance	Private sector	India, Indonesia, Pakistan, South Africa, USA, Vietnam
	Nitrogen use efficiency	Private sector	USA
	Abiotic stress and yield	Private sector	USA
Mustard	Male sterility	Private sector	India
Orange	Bacterial resistance	Private sector	USA
Pigeonpea	Insect resistance	Public sector	India

Continued

Table 5.1 Continued

Crop	Trait	Type of research institution	Countries
Potato	Fungal resistance	Public sector	Bangladesh, Indonesia, India
	Virus resistance	Public–private partnership	Argentina
	Various quality traits	Private sector	USA
Rice	Insect resistance	Public sector	China
	Insect resistance	Private sector	India
	Nitrogen use efficiency, water efficiency, salt tolerance	Public–private partnership	Colombia, Ghana, Nigeria, Uganda
	Nitrogen use efficiency	Private sector	USA
	Iron content	Public sector	India
	Provitamin A content	Public sector	Bangladesh, India, Indonesia, Philippines
	Stacked insect resistance and herbicide tolerance	Private sector	Argentina, USA
Safflower	High oleic acid	Public sector	Argentina, Australia
Sorghum	Stacked provitamin A, iron, zinc	Public–private partnership	Kenya, Nigeria
Soybean	Modified fatty acids	Private sector	USA
	Yield enhancement	Private sector	USA
	Multiple pest resistance	Private sector	USA
Sugarcane	Stacked insect resistance and herbicide tolerance	Private sector	Australia, USA
	Drought tolerance	Public–private partnership	Brazil, Indonesia
Tomato	Fungal resistance, insect resistance	Private sector	Argentina, Chile, Guatemala, India
	Fungal resistance, insect resistance	Public sector	China, Egypt
Wheat	Drought tolerance	Public sector	Australia, Egypt
	Insect resistance	Public sector	UK
	Fungal resistance	Public sector	China
	Virus resistance	Public sector	China
	Herbicide tolerance	Private sector	USA
	Improved grain quality	Public sector	Australia

Source: Own presentation with data from Raney and Matuschke (2011), James (2014), ISAAA (2015), GMO Compass (2015).

fungal, nematode, virus, and bacterial resistance. Various technologies with tolerance to abiotic stresses, such as drought and salinity, have also entered the field-trial stage in several countries. And finally, several GM output traits are already being tested, including biofortified crops with enhanced levels of micronutrients.

What are the expected effects of these new types of GM crops? Similar to Bt crops already on the market, pest and disease resistance will reduce chemical pesticide use and increase effective yields. As seen in chapter 4, the magnitude of these effects will be situation specific, depending on pest infestation levels and the effectiveness of pest control in conventional farming. In general, the yield effects of pest- and disease-resistant GM crops will be stronger in the tropics and subtropics, where infestation levels are often higher and farmers face more severe constraints in controlling crop damage. Especially under non-commercial smallholder conditions, where technical and economic constraints impede a more widespread use of chemical pesticides, crop losses are often 50 percent and higher (Oerke, 2006). Accordingly, the biggest yield gains are expected in South and Southeast Asia and sub-Saharan Africa (Table 5.2).

However, bigger expected yield gains in developing countries do not imply that pest- and disease-resistant crops cannot be of significant benefit also in developed countries. The example of virus-resistant papaya saving the Hawaiian papaya industry was discussed in chapter 4. Another relevant example with large benefit potential in terms of both pesticide reductions and higher effective yields is late-blight-resistant potato,

Table 5.2 Expected yield effects of pest- and disease-resistant GM crops in different regions

Region	Pest and disease pressure	Availability of chemical alternatives	Adoption of chemical alternatives	Expected yield effect of GM crops
Developed countries	Low to medium	High	High	Low
Latin America (commercial)	Medium	Medium	High	Low to medium
China	Medium	Medium	High	Low to medium
Latin America (non-commercial)	Medium	Low to medium	Low	Medium to high
South and Southeast Asia	High	Low to medium	Low to medium	High
Sub-Saharan Africa	High	Low	Low	High

Source: Modified and updated from Qaim and Zilberman (2003).

which is near to being commercialized in the United States, as was also discussed in chapter 4. Bacterial-resistant oranges are currently tested in Florida. This GM technology provides protection against citrus greening, a devastating bacterial disease that can hardly be controlled with conventional means. The resistance gene used was isolated from spinach. Resistance genes to citrus greening could not be identified in the orange genome. Moreover, orange trees are very difficult to cross-breed conventionally (Harmon, 2013).

The effects of GM crops with tolerance to abiotic stresses will also be situation specific. As the first commercial experience in the United States has shown, drought-tolerant GM varieties can lead to significantly higher yields than conventional varieties under water stress, whereas the effect is small when sufficient water is available. Drought-tolerant crops are of particular interest for arid and semi-arid environments. Many smallholder farmers in developing countries operate under drought-prone conditions. In a study covering eight low-income countries in Asia and sub-Saharan Africa, Kostandini et al. (2009) estimated that the average yield gains of GM drought tolerance may be 18 percent in maize, 25 percent in wheat, and 10 percent in rice. Drought tolerance will also contribute to higher yield stability (variance reduction), which is of particular relevance for risk-averse farmers that have no access to crop insurance. A few *ex ante* studies that were carried with partial and general equilibrium models suggest that the aggregate welfare gains of drought-tolerant crops could be very sizeable (Hareau et al., 2005; Kostandini et al., 2009; Qaim, 2009; Carter et al., 2011).

While the development of drought-tolerant and salt-tolerant varieties is a major priority both in public and private sector crop improvement programs, biotech researchers are also working on tolerance to other abiotic stress factors such as heat, flood, and coldness. Climate change is associated not only with rising average temperatures but also with more frequent weather extremes. Hence, more tolerant and hardier crops can help to reduce the risk of crop failures and food crises. Furthermore, research is underway to develop crops with higher nutrient use efficiency (see further details below). Some of these traits are genetically complex, so that commercialization may not be expected in the short run. Combinations of transgenic approaches with other tools of modern biotechnology and conventional breeding seem to be particularly promising to address the technical challenges. In the medium and long run, such new crop technologies could contribute remarkably to sustainable production increases and food security, especially in developing countries.

In the following sections, I look at the status and potential impacts of some of these new technologies more specifically with a particular focus on drought tolerance, nutrient use efficiency, and biofortification. Furthermore, I review recent developments in wheat, one of the most important food crops worldwide that has recently received renewed attention by biotechnology researchers. And finally, I focus on the situation and trends in Africa, where agricultural productivity is still very low but biotech capacities and interesting technologies are emerging in some of the countries.

Drought-Tolerant Crops

Drought, defined as an extended period when a region receives below-average precipitation, is a phenomenon that often has severe economic, social, and humanitarian consequences. Droughts can occur everywhere, but the consequences are usually much more severe in developing countries, due to insufficient capacity to deal with such emergencies. While in the twentieth century, repeated droughts had caused serious famines and mass starvation in China, India, and other countries of Asia, during the last 30 years human suffering related to droughts has been much more severe in Africa. The 1984–1985 drought in the Horn of Africa led to acute food shortages, killing an estimated 750,000 people and destroying the livelihoods of almost 10 million in Ethiopia, Somalia, Kenya, Uganda, and neighboring countries (von Braun et al., 1998). The same region has been afflicted by several severe droughts since then. Similarly, the Sahel region of West Africa is regularly hit by serious droughts with millions of poor people affected. Drought-tolerant crops alone will not be able to avoid food shortages and famine, but they can be an important component in strategies aimed at increasing the resilience of agricultural production and poor people's livelihoods.

In addition to acute droughts, increasing water scarcity is a significant challenge for global agricultural development. The agricultural sector already accounts for 70 percent of the global freshwater use. An estimated 20 percent of the world arable land is irrigated, producing 40 percent of total food. In many parts of the world, water withdrawal exceeds replenishment rates, so that groundwater tables are falling and rivers drying up. Climate change may be an additional challenge, as it is expected to increase water stress, especially in sub-Saharan Africa and South Asia. In sub-Saharan Africa, most agricultural production depends on rainfall, which is likely to become more erratic. In contrast, agriculture in large parts of South Asia is irrigated and depends on the melt waters from the Himalayas that regularly fill the rivers that are used to withdraw irrigation water. However, the Himalayan glaciers are shrinking, so the availability of irrigation water will likely decline in the decades to come. At the same time, demand for food and other agricultural commodities is rising rapidly. The development of agricultural systems that use water more efficiently will be a necessary condition for sustainable production increases. This will require improved agronomy combined with more water-efficient crop varieties (Reynolds, 2010).

Several public and private sector research organizations are working toward improving the water efficiency of important staple food crops, such as rice, wheat, and maize (Kempken and Jung, 2010). This involves both conventional breeding and genetic engineering. One public–private partnership project that has gained some publicity is WEMA. WEMA was started in 2008 and involves the International Maize and Wheat Improvement Center (CIMMYT), Monsanto, the African Agricultural Technology Foundation (AATF), and several other public sector institutes in Africa. This project is funded by the Bill and Melinda Gates Foundation and other donors. As part of WEMA, Monsanto and BASF donated the DroughtGard technology (event MON87460) for use in five African countries, namely Kenya, Mozambique, South Africa, Tanzania, and Uganda. Under the same project, Monsanto also agreed to donate its Bt maize technology (event MON810), which provides effective resistance to stemborer insect pests.

The drought-tolerance and insect-resistance traits were transferred to local African maize hybrids. GM hybrids with drought-tolerance and insect-resistance traits were already tested in confined field trials for various seasons in Kenya, South Africa, and Uganda. Field testing of hybrids with the stacked genes started in 2015. After further testing and commercial approval, stacked drought-tolerant and insect-resistant maize hybrids will be marketed royalty-free to smallholder farmers through local seed companies. Due to the country's experience with the commercialization of other GM crop technologies, South Africa is expected to be the first country to deploy WEMA hybrids from 2017 onward (James, 2014). Kenya and Uganda are likely to follow in subsequent years. In Mozambique and Tanzania, biosafety regulatory procedures first had to be established. In these two countries, the first confined field trials with GM maize may start in 2015. The goal of the WEMA Project is to increase maize yields by 20–35 percent over the yields of conventional hybrids during moderate drought years (CIMMYT, 2015). This is an important technology for Africa, given that an estimated three hundred million Africans depend on maize as their main staple food crop.

Similar projects to develop drought-tolerant crop varieties are also underway in other parts of the world, including in Asia. The International Rice Research Institute (IRRI) has recently released conventional drought-tolerant rice varieties in Bangladesh, Nepal, India, and other countries of Asia. These conventional varieties help to stabilize yields under drought conditions. However, breeding for abiotic stress tolerance with conventional approaches alone is a slow process. Significant efficiency gains are possible when combining conventional techniques with tools of genetic engineering (Mangrauthia et al., 2014). It needs to be clear that the release of the first drought-tolerant varieties is not the end to the research on improving the plant's water efficiency, neither in rice nor in any other crop. Continued research will be required to further improve the genetics and performance under water stress.

Related to water stress is the problem of soil salinity. At least 30 percent of the global irrigated land and 7 percent of the rainfed land are affected by salinity stress (Baulcombe et al., 2013). On some of these lands, salinization has made crop production impossible. Soil salinization can be reduced through better irrigation practices. However, rising sea levels and—related to this—increasing frequency of seawater intrusion may exacerbate the problem in coastal areas. GM salt-tolerant rice was developed and is currently tested in field trials. This technology is expected to increase grain yields on saline soils by 25 percent (Schroeder et al., 2013). The same technology could also be used in several other crops, such as soybean, wheat, and vegetables. Salt-tolerant varieties could also bind excess salt from the soil into the plant and thus help to rehabilitate salinized lands over time (James, 2013).

Nutrient Use Efficiency

Plants require various mineral nutrients for healthy growth and production. Limitation in any of the required nutrients reduces crop yield and quality and also makes the plant more susceptible to pests and diseases. In intensive agricultural production, fertilization with nitrogen, phosphorous, and potassium (NPK) is routine practice. Additional fertilization with micronutrients, such as iron, zinc, and selenium, is sometimes required depending on soil types and geology. Fertilization represents a significant economic and ecological cost. Mineral nitrogen fertilizer is produced from atmospheric nitrogen through the Haber-Bosch process. This process is very energy intensive and consumes close to 5 percent of the world's natural gas production. Phosphorous is a non-renewable resource, and minable rock phosphate reserves are finite (Cordell et al., 2009; Cordell and White, 2015). Fertilizers used in agricultural production also cause soil, water, and atmospheric pollution. A significant amount of nitrogen that cannot be absorbed by plants volatilizes as nitrous oxide, a greenhouse gas that is three hundred times more damaging than carbon dioxide.

When considering these environmental problems, a common reaction is to call for lower fertilizer intensity in agricultural production. However, in many situations lower fertilizer quantities will also mean lower yields, as comparisons between low-input and high-input systems clearly show. Reducing fertilizer use would therefore lead to higher land requirements to produce the same quantity of output. The problem is that the nutrient efficiency of crop plants is low. Average nitrogen use efficiency in cereals is below 50 percent, implying that the plants can only use less than half of the nitrogen fertilizers applied. Increasing nutrient use efficiency of plants through improved genetics would be a fundamental step toward more sustainable agricultural production.

Crop breeding during the twentieth century has focused on yield improvements, often without considering the constraints of fertilizer inputs. The high-yielding varieties developed and released during the Green Revolution required higher amounts of fertilizers, in order to perform successfully. Improving the nutrient use efficiency of plants was not an important breeding objective (Hawkesford, 2012). This has changed more recently with modern biotechnology and genetic engineering offering completely new perspectives. Various research groups and organizations are now focusing in particular on increasing nitrogen and phosphate use efficiencies (Kempken and Jung, 2010; Baulcombe et al., 2013).

In high-input production systems, plants with higher nutrient use efficiency will allow reductions in fertilizer use without jeopardizing yields. In low-input production systems, where plants suffer from insufficient nutrient availability, the same technologies could contribute to significant yield increases. Especially in the small farm sector of Africa the use of fertilizer is extremely low, due to knowledge, infrastructure, and financial constraints. Poor infrastructure conditions and limited market access were also important reasons why the Green Revolution did not take off in Africa to the same extent as in Asia and Latin America.

Arcadia Biosciences, a US company based in California, is one of the leading organizations concerning research on nitrogen use efficiency. Using genes for alanine aminotransferase from barley, Arcadia has successfully transformed canola and rice to increase nitrogen use efficiency. These GM crops were field-tested over various growing seasons in the United States, with yields similar or higher than those of the control varieties but two-thirds lower quantities of nitrogen fertilizer (James, 2014). Arcadia has many collaborative projects on nitrogen use efficiency with various private and public sector partners in developed and developing countries. This involves not just canola and rice but also several other crops such as maize and wheat. Some of these are commercial projects, while others involve technology donation for use by smallholder farmers in Africa.

One such project involving technology donations is the Nitrogen Use Efficient, Water Use Efficient, and Salt Tolerant (NEWEST) rice project involving Arcadia, AATF, the International Center for Tropical Agriculture (CIAT), the Public Intellectual Property Resource for Agriculture (PIPRA), and national agricultural research organizations in Ghana, Nigeria, and Uganda. This project was started in 2008 and is funded by USAID under the Feed the Future Initiative. Arcadia donated its nutrient use efficiency, water efficiency, and salt tolerance technologies to AATF for use in African NERICA type rice varieties under a royalty-free licensing agreement. The first field trials with nitrogen use efficient GM rice lines were carried out by CIAT in Colombia as an initial technology validation. After two seasons of testing with reduced nitrogen fertilizer applications, the project consortium reported that the new rice lines out-yielded conventional NERICA control varieties by 22–30 percent (Reuters, 2013). The same rice lines are now tested in confined field trials in Ghana, Nigeria, and Uganda. It may still take several years until the nitrogen use efficient varieties, and varieties with stacked nitrogen use efficiency, water efficiency, and salt tolerance will be commercialized in Africa. The benefits for smallholder farmers are expected to be substantial.

Arcadia's technology and many other initiatives on nitrogen use efficiency try to improve the genetics such that the plants are able to use more of the nitrogen that is available in the soil. A different mechanism is to improve the genetics such that the plants are able to fix their own nitrogen from the atmosphere. Legumes have this ability through their symbiotic interaction with nitrogen-fixing bacteria. Researchers are trying to extend this symbiotic interaction to a wider range of crops, particularly cereals. This involves both conventional and transgenic approaches (James, 2014). One large, international initiative funded by the Bill and Melinda Gates Foundation is the Engineering Nitrogen Symbiosis for Africa (ENSA) Project (ENSA, 2015). ENSA is led by the John Innes Center in the UK and also involves several other institutes in Europe and the United States. The project focuses on maize as the major staple food in Africa. However, engineering the nitrogen fixation capability into maize and other non-legume crops is a rather long-term objective and still involves significant basic research. Hence, this research runs in parallel to other initiatives to improve nitrogen use efficiency through breeding and improved agronomic practices.

Biofortified Crops

Micronutrient malnutrition is a widespread problem in many developing countries. An estimated two billion people are anemic, often due to iron deficiency. Over two billion people are at risk of zinc deficiency, and hundreds of millions lack one or more essential vitamins (WHO, 2015). The prevalence is especially high among the poor, whose diets are usually predominated by relatively cheap staple crops, with insufficient quantities of higher-value nutritious foods. Micronutrient deficiencies are often the cause for increased mortality and morbidity, so that the resulting health burden can be immense. For instance, vitamin A deficiency is a leading cause for child mortality and blindness in developing countries. Iron deficiency anemia contributes to maternal mortality and causes physical and cognitive impairment in children. In addition to the human suffering, such health problems also entail significant economic costs in the developing world (Stein and Qaim, 2007).

Controlling micronutrient malnutrition is regularly ranked as a top development priority (Lomborg, 2012). Economic growth and poverty reduction will help to reduce this problem in the long run, because richer people tend to diversify their diets toward more expensive and more nutritious foods. However, this is a slow process, so that more targeted micronutrient interventions are required in the meantime. Common micronutrient interventions include food supplementation (i.e., the distribution of micronutrient pills), industrial fortification (i.e., adding micronutrients to foods during industrial processing), and nutrition education programs. While all these interventions are very useful, regular implementation is costly. Moreover, these interventions do often not achieve wide coverage among the poor in rural areas (IFPRI, 2014).

Biofortification is an agriculture-based micronutrient intervention that involves breeding staple food crops for higher mineral and vitamin contents. The objective is to increase micronutrient intakes among the poor, thus improving their nutrition and health status. By focusing on staple crops that poor people heavily rely on, the approach is self-targeting. Biofortification could also be more sustainable than alternative micronutrient interventions. With a one-time research investment, biofortified germplasm can be shared internationally, and the varieties could spread through existing seed distribution systems. Since biofortified seeds can easily be reproduced, poor farmers in remote rural areas, with limited access to formal seed markets, could also be reached. Thus, unlike other micronutrient interventions, which require large funds on a regular basis, biofortification could produce a continuous stream of benefits with only small recurrent costs (Qaim et al., 2007).

Biofortification does not always involve genetic engineering. Several initiatives, including HarvestPlus, a research program of the CGIAR, primarily use conventional breeding techniques to increase micronutrient contents in wheat, maize, rice, sweetpotato, and several other crops. The first orange sweetpotatoes with elevated contents of provitamin A were recently released in Mozambique and other African countries (Hotz et al., 2012). However, with transgenic approaches higher levels of micronutrients can usually be achieved. Moreover, recombinant DNA techniques can help to introduce nutrients that are not found in the edible parts of certain crop species or their wild relatives. A case in point is provitamin A in rice. Finally, transgenic approaches are useful for crops that cannot be bred easily with traditional approaches, such as banana.

The most famous GM biofortified crop is Golden Rice, where several genes were introduced to produce provitamin A (beta-carotene) in the endosperm of the grain (see below for more details). Another example is Africa Biofortified Sorghum (ABS), a project that is jointly carried out by Pioneer/DuPont, Africa Harvest Biotech Foundation International, and national partners in various African countries. Pioneer has donated technologies to increase bioavailable iron and zinc contents for use in African sorghum varieties. Furthermore, ABS has used the Golden Rice technology also for sorghum transformation to increase provitamin A contents. GM sorghum varieties with these nutritional traits have been tested for several seasons in confined field trials in Kenya and Nigeria. Sorghum is an important staple food in several countries of West Africa.

Similar projects are underway to improve the contents of provitamin A in banana and cassava, with first field trials ongoing in Kenya and Uganda. Banana and cassava are of particular importance for poor subsistence farmers in East Africa. Using transgenic approaches, researchers have also managed to develop rice with high folate content, which will be combined with other nutritional traits to produce multi-biofortified rice (De Steur et al., 2012). Multi-biofortification is of interest, because many poor people suffer from multiple micronutrient deficiencies.

None of these GM biofortified crops has yet received commercial approval, so that *ex post* impact studies do not exist. There are several *ex ante* studies, however, that have tried to predict likely impacts, combining existing health and nutrition data with results from laboratory experiments and feeding studies. There are also studies that analyzed questions of consumer acceptance of GM biofortified foods (Lusk, 2003; González et al., 2009).

Zimmermann and Qaim (2004) and Qaim et al. (2007) developed a methodology to quantify the benefits and cost-effectiveness of biofortified crops in an *ex ante* framework. Since micronutrient malnutrition causes a significant health burden, which would be reduced through higher micronutrient intakes, they quantified the health burden with and without biofortification and interpreted the difference as the technological benefit. To make different types of health problems—such as premature deaths and various illnesses—comparable, the health burden is quantified through disability-adjusted life years (DALYs), a common measure in health economics. This methodology was used to assess the likely impacts of various biofortified crops in different countries (Qaim et al., 2007; De Steur et al., 2015). The results show that biofortified crops can contribute to large nutrition and health benefits with a high cost-effectiveness. However, the *ex ante* evidence also suggests that biofortification should not be seen as substitute for other micronutrient interventions, because all interventions have their strengths and weaknesses in particular situations.

Golden Rice

Golden Rice has been genetically engineered to produce beta-carotene (provitamin A), a precursor of vitamin A, in the endosperm of the grain. Conventional rice varieties do not have any beta-carotene; certain landraces contain tiny amounts in the outer layers of the grain, but not in the endosperm, which is what is usually consumed. Hence, traditional cross-breeding was not an option for the development of beta-carotene-rich rice. Golden Rice could potentially reduce vitamin A deficiency and associated health problems in rice-eating populations. Supporters of GM crops see Golden Rice as a compelling example of a pro-poor technology, whereas opponents have declared it a completely misguided approach. Golden Rice has become one of the centerpieces in the public controversy over GM crops.

Research on Golden Rice started in the 1990s by Ingo Potrykus at the Swiss Federal Institute of Technology (ETH) in Zurich and Peter Beyer at the University of Freiburg in Germany. At that time, hardly anyone believed that it would be possible to develop rice with beta-carotene. Yet, Potrykus and Beyer proved the sceptics wrong. In 2000, they published a paper in which they described how the transfer of genes from daffodils and bacteria helped to activate the beta-carotene biosynthetic pathway in the rice endosperm (Ye et al., 2000). As beta-carotene is orange, the transformed rice lines had a yellow-orange hue, which led to the nickname "Golden Rice." This proof of concept was hailed as a big scientific breakthrough internationally. In July 2000, Potrykus was pictured on the cover of Time Magazine with the headline "This rice could save a million kids a year." The initial quantity of beta-carotene was relatively low, with only 1.6 micrograms per gram of rice.

Further research work was required to increase the beta-carotene content in the rice. In 2001, Potrykus and Beyer negotiated a humanitarian licensing agreement with the company Syngenta. In a cashless transaction, they transferred the patents to Syngenta but retained the rights for any humanitarian use of the Golden Rice technology. Syngenta obtained the rights to use Golden Rice commercially, but agreed to support the humanitarian, non-profit visions of the inventors and their public sector licensees. (In 2015, the Golden Rice Project, and this innovative licensing agreement in particular, received the Patent for Humanity Award by the White House Office of Science and Technology Policy and the US Patent and Trademark Office.) Syngenta continued the research and managed to increase the quantity of beta-carotene in the rice endosperm to over 30 micrograms per gram by replacing the daffodil gene with an equivalent gene from maize (Paine et al., 2005). In 2004, the company stated that it had no interest in commercial exploitation of this technology.

Further work with the Golden Rice under the humanitarian license was carried out at IRRI in the Philippines. IRRI introduced the technology to Asian rice varieties, which were tested in the lab and in the field together with the Philippine Rice Research Institute (PhilRice). Golden Rice was also transferred to public sector research institutes in Bangladesh, India, Indonesia, and other countries in Asia. IRRI and PhilRice prepared the regulatory dossier for biosafety and food safety approval in the Philippines. This led to unexpected hurdles that delayed the project by several years and increased the cost of product development considerably (Potrykus, 2010). The antibiotic resistance marker genes had to be eliminated and numerous additional studies were required, many of them involving distinct and lengthy application and approval procedures. Approvals were often delayed, in some cases due to the direct influence of anti-biotech pressure groups. A field trial that cannot start on time can entail the loss of a whole growing season and the holdup of important follow-up work. Feeding trials, which are necessary to prove the technology's food safety, efficacy, and the bioavailability of the beta-carotene for humans, depend on the provision of sufficient quantities of Golden Rice. Such quantities can only be obtained from harvesting the field trial material. In 2013, a Golden Rice field trial was vandalized by activists in the Philippines, leading to the loss of some of the research data and material (Dubock, 2014).

In spite of all these hurdles, most of the studies have been completed, so that Golden Rice might possibly be approved for release to farmers in 2016 or 2017. The technology is likely to be commercialized first in the Philippines, with subsequent releases in other countries of Asia. Rice farmers in these countries will receive locally adapted rice varieties without having to pay a premium. Farmers will also be allowed to reproduce their own Golden Rice seeds. The rice is safe for human consumption and it will not taste different from normal rice. The beta-carotene in Golden Rice was shown to be highly bioavailable (Tang et al., 2012). Vitamin A toxicity through overconsumption of Golden Rice cannot occur because the human body converts only those amounts of beta-carotene to vitamin A that it needs, any excess supply is excreted. Whether farmers and consumers will accept the yellow-colored rice remains to be seen. Focus group discussions carried out in the Philippines suggest that the color may not be an issue (Zimmermann and Qaim, 2004). But there is the risk that opposing NGOs could disseminate negative propaganda against this yellow-colored rice, once it has been approved for commercial use.

With my research team I carried out an *ex ante* study on the potential impacts of Golden Rice in India (Stein et al., 2008). In India, mean levels of rice consumption are relatively high, and vitamin A deficiency is widespread. Of the 140 million pre-school children suffering from vitamin A deficiency worldwide, more than 35 million live in India. Our calculations show that the annual health burden of vitamin A deficiency in India is equivalent to a loss of 2.3 million DALYs, of which 2.0 million are lost due to child mortality alone. Widespread consumption of Golden Rice could reduce this burden by 59 percent, which includes the prevention of almost 40,000 child deaths each year (Qaim, 2010). The positive effects are most pronounced in the poorest income groups who suffer most from vitamin A deficiency and rely heavily on rice, as they cannot afford sufficient quantities of higher value foods.

We also calculated the expected cost-effectiveness by taking into account the actual and projected costs of developing, testing, and disseminating Golden Rice in India. Dividing the costs by the number of DALYs saved results in the average cost per DALY saved, which is a common measure for the cost-effectiveness of health interventions. According to our projections, the cost per DALY saved through Golden Rice is in a magnitude of 3 US$ (Qaim, 2010), which is very low. Usually, health interventions are considered very cost-effective when their cost is less than 200 US$ per DALY saved. These results show that Golden Rice is not a magic bullet to address vitamin A deficiency. But it could help to reduce the problem significantly and in a cost-effective way. Wesseler and Zilberman (2014) have calculated that every year of delayed approval of Golden Rice technology costs India at least 200 million US$ in terms of foregone social benefits.

Biotech Developments in Wheat

Wheat is one of the most important food crops worldwide, yet it has not received the same R&D attention as maize, soybean, canola, or other commercial crops. For the private sector, investments in wheat R&D are difficult, because farmers around the world often use farm-saved seeds, so that commercial seed sales are lower than for many other crops. A cost-effective hybridization technology for wheat has not yet been success-fully developed. For GM research, the fact that wheat is primarily used for human consumption further adds to the complexity. Consumers are much more concerned about GM technologies in typical food crops than about the same technologies in crops that are primarily used for feed and fiber production. This is the main reason why neither GM wheat nor GM rice have yet been approved for commercial use. Monsanto had devel-oped herbicide-tolerant wheat varieties, but the company dropped this research in 2004, because wheat farmers, millers, and traders feared for the stability of their export markets when adopting GM wheat (Raney and Matuschke, 2011). Concurrently, Syngenta was developing GM wheat with Fusarium resistance, but never pursued commercialization of this technology (Wilson, 2014).

Given the low investment in wheat and the fact that other major field crops experienced significant productivity gains through biotech and other research, the area grown with wheat in the United States and South America declined sharply, while the areas cultivated with maize and soybean increased. The relative profitability of wheat further decreased over time, so that in 2009 wheat farmers and stakeholders reached out to Monsanto and other biotech companies asking to resume the research work on GM wheat (James, 2014). The private sector decision to restart research on biotech wheat was also fueled by the 2008 food crisis, with international prices for wheat and other food grains rising significantly. In 2014, the International Wheat Yield Partnership (IWYP) was formally launched. IWYP is a consortium of public and private sector research organizations, research funders, and international aid agencies with the objective to stimulate new breeding research and make scientific discov-eries available to wheat farmers in developed and developing countries. The growing list of partners pledged to invest 100 million US$ over the first five years (IWYP, 2015). The IWYP consortium aims at increasing wheat yields by 50 percent in the next 20 years.

Research under IWYP and other breeding work in wheat does not build on recombinant DNA techniques alone. But genetic engineering plays an important role, so it is likely that GM wheat will be released sometime in the next 5 to 10 years. Herbicide-tolerant wheat varieties will likely be the first to be commercialized. Other GM traits under development in wheat include insect, fungal, and virus resistance, drought tolerance, nitrogen use efficiency, and various quality traits (James, 2014). Some of these technologies are already tested in the field (see Table 5.1). Field trials with drought-tolerant GM wheat varieties carried out in Australia showed that yield gains of 20 percent are possible under drought-prone conditions (Raney and Matuschke, 2011; Wilson, 2014). Other important areas of research are physiological traits for improving heat tolerance in wheat. The worst impacts of climate change and rising temperatures will likely occur at low altitudes, where 100 million hectares of wheat are currently cultivated (Cossani and Reynolds, 2012).

IWYP is a consortium of public and private research organizations, and much of the research involving plant genetic engineering in wheat is carried out through public–private partnerships. This is in part because most of the wheat germplasm is controlled by the public sector, while most of the GM traits of potential interest in wheat are being developed by private sector companies (Wilson, 2014). The development and commercialization of successful GM wheat varieties require the combination of both, superior and well adapted germplasm and interesting new traits.

Situation and Trends in Africa

From an agricultural development perspective, the situation in Africa deserves particular attention. While Asia is the region with the largest absolute number of poor and undernourished people, poverty rates relative to the total population are much higher in Africa. In sub-Saharan Africa, about one-quarter of the population still suffers from insufficient calorie intakes (FAO, 2015a). Micronutrient malnutrition is even much more widespread. The majority of the poor and undernourished people in Africa depend directly on agriculture as a source of income and employment. Africa is also the region where population growth is highest. The population in Africa is projected to more than double from less than one billion today to 2.4 billion in 2050 (United Nations, 2014). Against this background, increasing agricultural productivity has to be one of the top priorities to reduce poverty and foster sustainable growth and food security in Africa. Given that a large proportion of the poor are smallholder farmers, this productivity growth has to take place in the small farm sector. This does not mean that structural change should not happen over time. But this needs to be a gradual process, driven by rural people finding more lucrative employment in other sectors rather than being pushed out of agriculture due to destitution and lacking prospects for improving their livelihoods.

Historically, technological developments in agriculture were much less successful in Africa than elsewhere. The input-intensive model of the Green Revolution, which has boosted agricultural production in Asia since the 1960s, was not widely adopted in Africa due to various constraints (Eicher and Staatz, 1998). Accordingly, yield levels for most crops in Africa are far below the average yields obtained by farmers in Asia or South America, where the climatic conditions are often similar (Table 5.3). For many crops, yield gaps in Africa are 50 percent and higher. Productivity increases cannot be achieved through new technologies alone. The experience shows that improvements in infrastructure, more secure property rights over land, better extension and credit systems, and better governance are all crucial ingredients for broad-based productivity growth and rural development. But better and more suitable technologies will also have to play an important role. Combined with other technical and institutional innovations, GM crops offer large potentials for Africa, as they can help to increase yields sustainably, especially in situations where low amounts of other external inputs are used.

Okeno et al. (2013) have reviewed several *ex ante* studies, showing that even the existing GM technologies—such as Bt maize, Bt cotton, and Bt eggplant—could produce large benefits for African producers, consumers, and for the environment if they were approved locally. The benefits of emerging GM technologies—such as the drought-tolerant, nitrogen use efficient, and biofortified crops that are specifically developed for Africa (see above)—could even be much higher. So far, however, only Burkina Faso, South Africa, and Sudan grow GM crops commercially.

Table 5.3 Average yields of major crops in different world regions (kg/ha, 2012)

	Africa	Asia	South America	World
Banana	10,623	26,796	20,072	20,591
Cotton	1,000	2,742	2,717	2,231
Maize	2,057	5,168	4,820	4,888
Rice	2,623	6,301	5,100	4,548
Sorghum	1,009	3,366	3,596	1,495
Soybean	1,203	1,573	2,422	2,298
Wheat	2,416	3,028	2,558	3,090

Source: Own presentation with data from FAO (2015).

South Africa was an early adopter of GM crops and has grown Bt maize, Bt cotton, and HT soybean for many years. In Burkina Faso and Sudan, Bt cotton is the only GM crop that was approved for commercial use much more recently (Falck-Zepeda et al., 2013).

What are reasons for the slow uptake of GM technologies in Africa in spite of the large potentials? Low biotech capacities in most of the local research organizations and inadequate public investments in upgrading these capacities are important factors. Much more constraining, however, are the European-style regulatory systems for the approval of GM crop research, testing, and commercialization. Unlike in the United States or Canada, where GM crops are regulated by the same authorities that also regulate other agricultural and food technologies, in the European Union new laws and institutions were established for GM crop regulation. The development of biotechnology policies and the establishment of new regulatory bodies and institutions require political support for GM crops that was not always given. Observing the opposition against GM crops in Europe, many African policymakers were rather hesitant to support GM crops. Listening to the fabricated stories about "Frankenfood," many Africans believe that GMOs are a high-risk technology that would damage indigenous biodiversity and put nutrition and health of the local population at risk. African policymaking elites were often educated in Europe. Many also send their children to European universities and have tighter business and trade relationships with Europe than with any other part of the world. Hence, these elites are influenced significantly by the highly precautious European attitudes and regulatory approaches for GMOs, even though the circumstances in Africa are so different from those in Europe (Paarlberg, 2008). Also the media coverage about GM crops in Africa follows the European example and is much more focused on reporting potential risks.

One example from Kenya is quite characteristic. In 1999, a project to test Bt maize with resistance to maize stemborers was started by CIMMYT in cooperation with the Kenya Agricultural Research Institute (KARI). Under existing regulations, Bt maize varieties were field tested. These Bt varieties showed effective resistance against some of the major insect pests and were found to present no new risks for human and animal health or the environment. However, in 2004 it was decided that GMOs could not be released without a specific biosafety law. It then took five years to move from a draft law to a final act of parliament. The biosafety law was passed in 2009. It took additional three years to draft and publish specific guidelines for handling GMOs under this new law and to establish a functional national biosafety committee (Paarlberg, 2014). The Bt maize varieties have not been commercially approved up till now. Similarly, Bt cotton varieties, which had been tested in Kenya since 2004, have not been approved commercially. These are the same technologies that have been used successfully elsewhere in the world for many years.

More recently, Kenyan politicians have shown more support for GMOs. Similar trends are also observed in a few other African countries. Several countries in Africa recently put new biosafety policies in place, have upgraded their biotechnology capacities, and are welcoming to international initiatives focusing on GM crop development. In addition to the countries that have already commercialized GMOs, African countries that are now significantly involved in GM crop development and field testing include Cameroon, Egypt, Ghana, Kenya, Malawi, Nigeria, and Uganda. The key crops at various stages of experimentation in confined and open field trials in Africa include banana, cassava, cotton, cowpea, maize, rice, sorghum, sweetpotato, and wheat (James, 2014).

In addition to the technologies that were already discussed earlier, several other GM crops in the research and testing pipeline could be of particular relevance for smallholder farmers in Africa. Research groups in South Africa have developed GM maize with resistance to the maize streak virus, a pathogen that is indigenous to Africa and contributes to significant crop losses. The Donald Danforth Plant Research Institute in the United States has developed cassava with an inbuilt resistance mechanism against the cassava mosaic virus, which is currently field-tested under confined conditions in Kenya and Uganda. The Commonwealth Scientific and Industrial Research Organization (CSIRO) in Australia has developed Bt cowpea with resistance to the cowpea borer, a significant insect pest in Africa. Cowpea is the most important food legume in the semi-arid tropics and is consumed by close to two hundred million people in Africa. With support from AATF, Bt cowpeas are currently field-tested in several countries of West Africa. Finally, GM technologies to protect banana against weevils and bacterial blight are currently tested in Uganda. Research to develop GM banana with resistance to Black Sigatoka, a fungal leaf spot disease, is ongoing (Thomson, 2014).

These GM technologies could significantly increase and stabilize yields and incomes in the African small farm sector, without the need to increase the use of other external inputs. They also have the potential to improve food and nutrition security among farmers and consumers. Most of the GM technologies currently developed for Africa involve public sector research organizations and have humanitarian objectives. It is important to test these technologies for their effectiveness and safety, but unnecessarily delaying or blocking their release must be avoided, as the costs in terms of foregone social benefits could be large.

CHAPTER 6

GM CROP REGULATION

GMOs are more heavily regulated than any other agricultural technology. The regulation focuses primarily on the assessment and management of biosafety and food safety risks. Other important areas of regulation include labeling requirements for GM foods, as observed in some countries, and, related to labeling, rules of coexistence to facilitate segregation of supply chains for GM, conventional, and organic crops. Another area of regulation with immediate relevance for GM crops are intellectual property rights (IPRs) for biological materials and technologies. In this chapter, I review regulatory approaches and discuss the wider implications for GMO research, commercialization, international agricultural trade, and market structure in the biotech industry.

Biosafety Regulation

The roots of specific biosafety regulations for GMOs can be traced back to the Asilomar Conference on Recombinant DNA that was held in 1975 in Asilomar, California. This conference was organized by Paul Berg, an American biochemist and one of the first researchers to conduct studies with recombinant DNA. In the early 1970s, Paul Berg had combined fragments of different types of viruses when he realized that this research might possibly be associated with particular risks. The Asilomar Conference brought together around 140 researchers who discussed potential risks and drafted voluntary guidelines to ensure the safety of recombinant DNA technology. These voluntary guidelines significantly influenced the design of biosafety policies and regulations in many countries.

Ever since the Asilomar Conference, regulatory policies related to GMOs have focused on the idea that this technology poses new risks to human health and the environment (Fagerström et al., 2012). In addition to containment rules for recombinant DNA material in the laboratory, regulations and complex approval procedures for testing GM crops in greenhouses and in the field, and for commercially releasing GMOs were widely established. The notion that GMOs are inherently risky was further fueled by the Cartagena Protocol on Biosafety, which was adopted in Montreal in January 2000 after several rounds of negotiation between the UN member states. The Cartagena Protocol is a supplement to the UN Convention on Biological Diversity (CBD) signed in Rio de Janeiro in 1992. In the Cartagena Protocol, GMOs are singled out as a potential risk to genetic resources. The Protocol requires specific measures to reduce these risks and a special information and decision-making process when GMOs are to be exported to another country. The Cartagena Protocol also implies that policies related to GMOs should follow the precautionary principle, meaning that caution should be practiced in the context of uncertainty. The Cartagena Protocol on Biosafety came into force in September 2003 once it had been ratified by 50 states. The United States, Canada, Australia, Argentina, and several other countries that have commercialized GM crops have not ratified the Cartagena Protocol (Smyth et al., 2014).

In the 1980s and 1990s, GM crops were a relatively new phenomenon and there was substantial uncertainty related to the potential health and environmental risks. Hence, some precaution and the establishment of specific regulations to carefully test and reduce potential risks were justified. This situation has changed however. Twenty years of commercial experience with GM crops, 30 years of research, and hundreds of millions of dollars spent on risk assessment have shown that GM crops are not *per se* riskier than conventionally bred crops (EU Commission, 2010b; EASAC, 2013; House of Commons, 2015). This is not just the opinion of a small group of biotech researchers, but the conclusion drawn by a large number of organizations, including the World Health Organization (WHO), the Food and Agriculture Organization (FAO), the Organization for Economic Cooperation and Development (OECD), and numerous Medical Associations and Science Academies in the United States, UK, Germany, France, China, India, and several other countries (Dubock, 2014).

The conclusion that GM crops are not *per se* riskier than conventionally bred crops does not mean that all GM crops are completely free of any risk. For example, a hypothetical crop engineered to produce hydrocyanic acid would be extremely toxic to humans, but the same would hold true if a conventional variety were bred to produce hydrocyanic acid. Also for actually existing GM crops certain risks exist, as was discussed in chapters 3 and 4. Herbicide-tolerant GM crops—if used inappropriately—can encourage monocultures and buildup of resistance in weed populations, potentially fostering a pesticide treadmill. But the same risks occur with conventionally bred crops that are tolerant to non-selective herbicides. Such conventional herbicide-tolerant varieties were developed for certain crop species, and they are used without requiring any biosafety approval and without much public opposition (Tan et al., 2005).

This discussion shows that there are certain risks, so that some regulation is required. But the risks are related to the plant traits (i.e., the *product*), not the breeding approach (i.e., the *process*). Hence, it makes sense to regulate on a product basis rather than singling out one particular breeding approach as particularly risky. In fact, there is no scientific justification for a process-based regulatory approach with very different standards and procedures for GM and conventionally bred crops. Yet, following the Cartagena Protocol many countries actually use a process-based approach to regulate GM crops.

Differences in Regulatory Approaches

Risk assessment and risk analysis of GMOs is governed by internationally accepted guidelines, developed by the Codex Alimentarius of WHO and FAO. A leading principle of the Codex Alimentarius is the concept of substantial equivalence, which stipulates that any new GM crop technology should be assessed for its safety by comparing it with an equivalent, conventionally bred variety that has an established history of safe use (Fagerström et al., 2012). Despite the fact that the Codex guidelines are globally endorsed, significant differences in the GMO regulatory approaches can be observed between countries. The differences between the European and American approaches are particularly pronounced. While the European Union (EU) ratified the Cartagena Protocol with its specific rules for GMOs, the Protocol was not ratified by the United States, Canada, and most other countries in the Americas.

The regulatory approach in the EU requires new laws that are specific to GM crops and foods, while in the United States GMOs are regulated under the same laws that are also used for conventional agricultural technologies. Related to this, the EU approach also requires a separate testing and approval process for GM crops that is overseen by institutions especially established for this purpose. In the United States, existing institutions that also regulate conventional crops—namely the Food and Drug Administration (FDA), the Animal and Plant Health Inspection Service (APHIS), and the Environmental Protection Agency (EPA)—screen and approve GMOs. Finally, following the precautionary principle, even without any evidence of risk EU regulators can refuse to approve GM crops on grounds of uncertainty alone. In the United States, the pre-cautionary principle is not applied. If the required tests for known risks such as toxicity, allergenicity, environmental invasiveness, and effects on non-target organisms, have been passed successfully, there is no further regulatory hurdle for commercialization of the GM crop in question (Just et al., 2006; Paarlberg, 2014).

The regulatory approach for GMOs in Europe is clearly process based. The process of plant genetic engineering is singled out with rules and standards that do not apply to any other breeding method. In contrast, in the United States a product-based approach is followed. Legislation concentrates on the risks of the product—that is, the crop plant with a specific new trait—and not the breeding method, as genetic engineer-ing is not considered inherently more risky than conventional methods (Devos et al., 2010).

Reasons for the differences in the regulatory approaches between Europe and the United States are manifold. One important factor is that in Europe NGOs have enjoyed greater influence on policymaking pro-cesses since the mid-1980s (Ammann, 2014). In addition, differences in farming conditions and consumer attitudes play a role. The first GM crops that were commercialized in the United States in the mid-1990s were herbicide-tolerant soybeans, Bt cotton, and Bt maize. These tech-nologies are of great value for American farmers, but less so for their colleagues in Western Europe. Farmers in the EU hardly grow any cot-ton and soybean. They do grow maize, but the first Bt maize varieties provided resistance to insect pest species that are not yet a great problem in most parts of Europe. Hence, European farmers did not constitute a strong lobby in terms of protesting against exaggerated regulatory hur-dles for GMOs. Future GM crop technologies, and also a few existing ones such as herbicide-tolerant canola and sugarbeet, could be of greater value also for farmers in Europe.

Consumer preferences are heterogeneous not only between but also within Europe and the United States. But on average, Europeans tend to have lower trust in the authorities that regulate food. Europeans experi-enced a few regulatory failures in the 1980s and 1990s that undermined public confidence in the ability of regulatory officials to adequately pro-tect the public's health and safety. One such regulatory failure was mad cow disease (bovine spongiform encephalopathy, BSE). BSE is unrelated to GMOs but it still influenced public attitudes toward biotechnology. For many years, the EU Commission had declared that BSE would pose no danger to humans. This statement had to be revoked in the mid-1990s when there was growing evidence that exposure to cattle infected with BSE could contribute to Creutzfeldt-Jakob disease in humans. Around the same time, Europe also experienced several other food scares, including outbreaks of e-coli, salmonella, listeria, and foot-and-mouth disease, as well as the contamination of various foods with dioxin (Lynch and Vogel, 2001). Some of these cases represented real health issues, others were artificially scandalized. In any case, these incidents undermined the European consumer confidence in the safety of their food and the trustworthiness of the food regulatory authorities.

Beyond these food scandals and regulatory failures, differences in public attitudes toward food, agriculture, and the private sector may also have played a role for unlike regulatory approaches in Europe and the United States. So far, most GM crops were developed by private multinationals, which are generally considered with greater distrust in Europe than in the United States.

Regardless of the underlying reasons for regulatory differences, it is becoming increasingly clear that the European approach is stifling innovation in the broader area of plant sciences and is banning GM crops rather than allowing their safe use. So far, only two GM crop technologies were approved in the EU, compared to over 160 in the United States and close to 40 in Brazil (James, 2014). For many of the GM technologies that were approved in the United States, approvals were also sought in the EU, either for import or for commercial cultivation, but these applications were either denied or the cases are still pending with multiple years of delay. Several applications for GMO cultivation in Europe were also withdrawn by the applicants after years of stalemate in the approval procedure. The safety record for the GM crops approved in the United States and elsewhere is essentially unblemished. Hence, European regulators are repeatedly committing type II errors, meaning that safe technologies are not approved for commercial use in the EU.

Problems with the European Regulatory Approach

Many independent researchers have stated that the EU regulatory system for GMOs is not fit for purpose (House of Commons, 2015). The process-based regulatory approach has two fundamental flaws. First, it builds on the underlying premise that GM crops present higher and completely new risks as compared to their conventionally bred counterparts. While uncertainty with respect to this issue was still larger in the 1980s and 1990s, when the regulatory system was designed, 30 years of additional experience clearly suggest that this premise of new inherent risks of GMOs is unfounded. Second, process-based regulation assumes that there is one particular new breeding technique that is completely different from all others. As was discussed in chapter 2, there are several techniques that are counted as conventional breeding, although they can result in genome disruptions that are more profound and unpredictable than those of genetic engineering. A case in point is mutagenesis induced through radiation or chemical agents. Unlike GM crops, mutant varieties are not subject to any special regulation; they are even widely used in organic agriculture. I also discussed conventional breeding techniques that can produce "transgenic" varieties in the sense that they contain genes from foreign species (chapter 2). On the other hand, not all varieties resulting from genetic engineering are transgenic. As described in chapter 3, genetic engineering can also be used to produce cisgenic plants, which do not contain any foreign genes. And new genome editing techniques—such as TALENs or CRISPR/Cas-based procedures—are much more precise than the more classical approaches of plant transformation.

The EU's process-based regulatory system is increasingly proving itself incapable of dealing with these technological dynamics. Without a fundamental reform, new plant biotechnologies—whether transgenic or otherwise—will increasingly be hindered or halted, while potentially harmful conventional crops may escape responsible control (House of Commons, 2015).

However, the process-based approach is not the only problem with the European regulatory system. Other issues are related to the responsibilities of actually making GMO approval decisions. Any party seeking approval for a GM crop must provide a regulatory dossier including extensive scientific documentation demonstrating that the crop has no adverse effects on human health, animal health, and the environment. Putting together such a dossier requires several years of risk studies in the lab, in greenhouses, and through field testing. In the EU, regulatory dossiers are evaluated by the GMO Panel of the European Food Safety Authority (EFSA), which consists of independent scientific experts. EFSA was established in 2002. Before that, various national and EU-level institutions had different responsibilities. The establishment of a European-wide, independent risk assessment body was an important step toward increasing consumer confidence in the safety of food and food-related technologies.

However, approval decisions are not made by EFSA. EFSA provides a scientific opinion to the European Commission, which prepares a proposal for or against authorization. This proposal is then discussed in the Commission's Standing Committee for Food and Animal Health consisting of member country representatives. If this Standing Committee accepts the proposal, it is finally adopted by the Commission. Alternatively, it is passed on to the Council of Agricultural Ministers. If a qualified majority for or against the proposal cannot be reached in the

Council, which happens regularly due to the political meddling, the case is referred back to the EU Commission, which then adopts the proposal (Davison, 2010).

The opinion of EFSA's GMO Panel should form the scientific basis for the final decision in the EU Commission and the Council of Ministers. By that logic one might expect that only GMOs that were found to be associated with adverse health or environmental effects during testing would not be approved. In practice, however, the decisions are far from being based on scientific evidence alone. As a matter of principle, several EU member states vote against the approval of any GMO, regardless of the scientific opinion provided by EFSA (Fagerström et al., 2012; Dubock, 2014). The whole process is highly politicized with significant lobbying activities by all sorts of interest groups at national and EU levels. Many times, the precautionary principle is misused by coming up with new hypothetical risks, which may be farfetched but still hard to disprove formally. Science cannot demonstrate freedom from risk, especially not from unknown risks because absence of evidence is not the same as evidence of absence. Anyone who is opposed to GMOs for whatever reason can invent a hypothetical risk and thus contribute to a ban, or at least a substantial delay in the approval process.

One example related to GM maize vividly demonstrates the regulatory hurdles for GMOs in the EU. In 2000, Pioneer/DuPont submitted an application for the approval of maize 1507, a GM crop with stacked herbicide-tolerance and insect-resistance traits. At that time, the EU had an unofficial moratorium on GMOs. Although a different Bt maize event (MON810) had been approved before the moratorium started in 1999, any new approvals were suspended. The moratorium ended in 2003, after the establishment of EFSA as the new authority for independent risk assessment. EFSA then asked Pioneer/DuPont for additional data, which were provided. Based on these data, EFSA issued a positive safety opinion in early 2005. In a meeting with the European Commission and member countries additional concerns were raised. In 2006, the Commission asked EFSA to provide a further scientific opinion. In subsequent years, maize 1507 was assessed and reassessed multiple times, receiving a total of seven positive EFSA opinions (GMO Compass, 2015). According to EU laws and regulations, the Commission has to prepare an authorization document within three months of receiving an opinion from EFSA, but it repeatedly failed to do so. In 2010, Pioneer/DuPont filed a legal action against the Commission, which was successful. In September 2013, the General Court of the EU ruled that the Commission had failed to meet its obligations. Soon thereafter, the Commission passed a proposal in favor of approving maize 1507. However, in February 2014 the European

Parliament passed a resolution calling on the Council of Ministers to reject the Commission proposal on the basis that the long-term effects of the GM maize on non-target organisms had not been taken into account sufficiently (House of Commons, 2015). A final authorization decision on maize 1507 in the EU is still pending (as of mid-2015). In the United States, maize 1507 was commercially approved in 2001 and has been used there since then without any negative impacts.

There are other examples of GM crop technologies that took a similar fate in the EU. In 1996, BASF had handed in an application for its Amflora potato, a product with a modified starch structure to make it more useful for industrial purposes. Amflora also received multiple positive opinions from EFSA, but the approval process was repeatedly delayed. In 2010, the EU Commission eventually approved Amflora for feed and industrial purposes. After maize MON810 that was approved in the EU in 1998, Amflora became the second commercialized GM crop 12 years later. However, protests by anti-biotech groups continued and fields for seed production of Amflora potato were vandalized. In parallel, NGOs initiated legal action against the Commission's approval. Supported by several EU member countries, including particularly those with strong opposition against GMOs, a lawsuit was filed at the General Court of the European Union. In 2012, BASF decided to stop marketing Amflora potato in Europe. In the same year, the company also decided to completely move its plant science research from Europe to the United States. In 2013, the EU Court annulled the approval decision for Amflora in the EU due to a procedural error in the approval process (General Court of the European Union, 2013). The Court ruling said that the Commission's decision had been made without sufficiently involving member countries through the Standing Committee. In spite of the EFSA statements on the safety of Amflora for human health and the environment, NGOs depicted the EU Court's decision as being based on significant scientific concerns. Also in 2013, BASF decided to discontinue the pursuit of regulatory approvals for three additional GM potato technologies, including the Fortuna potato with late blight resistance (James, 2014).

However, even after authorization of a GMO by the European Commission, EU member countries can ban individual GM crops under a special safeguard clause, if there are "justifiable reasons" that the particular variety may cause harm to humans or the environment. The member country must then provide sufficient evidence that this is the case. Sufficient evidence, however, is ambiguous language. Several countries—including Austria, France, Germany, and Poland—used this safeguard clause to ban Monsanto's maize MON810. The German Minister of Agriculture banned MON810 in 2009; this was federal election year and the ban was a popular move in the German public. Old studies showing that Bt pollen could harm non-target insects under artificial laboratory conditions were submitted as "justifiable reasons." The German Central Commission for Biological Safety (ZKBS) considered the ban to be scientifically unfounded. Over 1,600 scientists had appealed to the German Minister not to sacrifice a future technology with great potential simply because of short-sighted political interest (Davison, 2010). The ban on MON810 in Germany and several other EU countries is still in place.

Policymakers in Europe often argue that they cannot ignore in their decision-making the fact that the majority of European citizens is against the use of GMOs. This is how regulatory hurdles and bans of GMOs are often justified unofficially. However, regulatory systems and ad-hoc decisions are not only a response to public attitudes, but they also contribute to forming public attitudes in a significant way. The process-based regulatory approach builds on the notion that GM crops are particularly risky. This notion is reinforced by the EU Commission and member country's habit to regularly ignore EFSA opinions and delay approvals on the basis of uncertainty. Also the individual-country bans of GM crops under the safeguard clause generate anxiety because the only avenue of preventing the cultivation of GM crops in the home territory is through arguments of risk, even though the real motivation for the ban is often totally unrelated to food safety or biosafety concerns.

Only very recently, the EU Commission agreed that individual member countries shall be free to choose whether or not to approve in their own territory the planting of GM crops that were judged safe by EFSA. This GMO opt-out clause was passed in January 2015 by the European Parliament. It allows member countries to ban GMOs on grounds other than risks to health and the environment. Such other reasons may be related to concerns about socioeconomic impact or certain farm policy objectives (European Parliament, 2015). It remains to be seen how this new legislation will affect regulatory processes.

On the one hand, the GMO opt-out clause could speed up the scientific approval process and reduce political lobbying at the EU level. This would allow those EU member countries with more positive attitudes toward new plant science technologies—such as the UK and Spain—to go ahead with the planting of GM crops. Separating the health and environmental safety approval at EU level, and the socioeconomic and policy arguments at national levels, could also contribute to a more differentiated public debate and understanding. It is noteworthy that anti-biotech groups were lobbying against this opt-out clause by EU member countries, fearing that it could alter the stalemate of GMO approvals in

Brussels that they had helped so carefully to put in place and maintain over many years.

On the other hand, several EU member countries now consider to ban GMOs permanently on their territories under the opt-out clause, even for testing purposes. Permanent bans in individual member countries would not only contradict the idea of a EU-wide free market, but could also entail further obstacles for international progress in plant breeding research. A permanent ban on GMO cultivation in a country would also mean a stop for any GMO research in that country. Europe has still some of the world's leading public sector research groups on plant genetic engineering. Some of the basic technologies of plant transformation were developed in European labs in the 1970s and 1980s.

Wider Consequences of Overregulation

Overregulation occurs when the marginal costs of regulation exceed the marginal benefits in terms of higher safety levels. This is clearly the case for GMO regulation in Europe, which has blocked the commercial release of technologies that were declared safe by EFSA. Kalaitzandonakes et al. (2007) estimated the compliance costs for the regulatory approval of a new GM crop to be in a magnitude of 6–15 million US$ in one country. These are the costs for the food safety and biosafety studies that need to be conducted in order to compile the regulatory dossier. The costs are fully borne by the applicant, that is, the innovating company or organization. Kalaitzandonakes et al. (2007) compiled and used data for a significant number of applications in different countries. However, naturally only cases of GM crops that were eventually approved could be considered. Most of these cases refer to the United States, Canada, Argentina, and Australia, where regulatory procedures are relatively efficient. While no concrete estimates are available, it is clear that the regulatory compliance costs are substantially higher in the EU. Every unexpected test that becomes necessary because of additional data requests, and every additional year of delay add to the costs and to the uncertainty whether the product will ever be approved. In several cases, companies have spent well over 20 million US$ on a single application without getting the technology approved.

These costs of complying with the regulatory procedures do not include the value of the foregone benefit, which can be substantial. Fagerström et al. (2012) calculated that not approving GM canola and sugarbeet with herbicide tolerance and GM potato with late blight resistance costs the EU about 2 billion US$ per year in terms of foregone benefits. That European farmers and consumers are hardly able to benefit from GM crops is unfortunate, but not a huge social problem, because people in the EU are relatively well-off. Farmers in the EU also receive significant political support through subsidies and government transfers. The much bigger problem is that Europe has exported its regulatory system also to other parts of the world, notably to Africa, where poverty and undernutrition are widespread.

The non-approval of GMOs and the overly precautious attitudes have much larger social costs in Africa in terms of lost opportunities to improve productivity and income in the small farm sector, as well as food security and nutrition in the wider population. The European influence on GMO regulatory systems in Africa occurs through various channels (Juma, 2011). Direct influence was exercised through bilateral foreign assistance. European governments used their development aid agencies to encourage African governments to establish European-style regulatory systems and also provided related technical support (Paarlberg, 2008). This was fostered by a multilateral project to develop national biosafety frameworks supported by the United Nations Environment Program (UNEP) to help African countries implement the Cartagena Protocol. UNEP is one of the UN organizations where the European influence is particularly strong. A more indirect channel of influence on policymakers in Africa is through the media. People in Africa follow the public debate about GMOs and the politicized process of regulatory approval in the EU quite closely, often through reports and interpretations in local newspapers. I was traveling in Africa in 2009 when the German Minister of Agriculture used the EU safeguard clause to ban Bt maize MON810 from cultivation due to "new concerns about impacts on beneficial insects." Several articles in African newspapers took this up, linking the GM maize to cancer and other dangers for human health.

In addition to its high costs in terms of compliance and foregone benefits, overregulation also affects directions of technological development and the structure of the biotech industry. Regulatory costs are now much higher than the actual cost of developing GM crops. And, regulatory costs accrue again for every new GM crop and in each additional country where this crop shall be commercialized. This introduces a bias in favor of large, commercial crops and countries with large agricultural areas. For local crops, such as millet, cassava, or sweetsspotato, and in small countries the markets are simply not large enough to justify the high fixed cost regulatory investments (Qaim, 2009). Furthermore, expensive and uncertain regulatory procedures are difficult to handle by small firms and public sector organizations, so these hurdles contribute to further concentration of the agricultural biotech industry. While multinational companies have the financial capacity to deal with this situation, or to move their activities to other places, many smaller companies and public universities in Europe have stopped developing and testing GM crops (Ricroch and Hénard-Damave, 2015). Even for large international research consortia that develop new crop technologies for humanitarian purposes the regulatory costs and hurdles are extremely challenging (Potrykus, 2010). Donor organizations are often willing to provide a few million dollars for the development of promising technologies targeted at poor farmers and consumers. However, finding organizations that are willing to fund additional tens of millions of dollars for the regulatory procedure in one single country has proven much more difficult.

GM Food Labeling and Coexistence

Several countries have introduced or considered introducing a food labeling system for GMOs, which is also directly related to public attitudes and consumer concerns. In general, labeling is used to inform consumers about ingredients, nutritive values, quality standards, health and environmental effects, and other non-visible product or process attributes of foods and other consumer goods. Depending on the type of effect, either mandatory or voluntary labeling is possible. Mandatory labeling is often used to warn consumers of specific health risks (e.g., cigarettes, allergenic potential), whereas voluntary labeling is more common to differentiate products with desirable characteristics for marketing purposes (e.g., organic, animal welfare). Both mandatory and voluntary labeling systems can convey the same information to consumers. Given that only GM products that are considered to be safe by the regulatory authorities are approved for commercialization, no warning of risks is required on labels. Therefore, the issue is mainly one of heterogeneous consumer preferences, which—from an economics perspective—would be best addressed through voluntary labeling of GM-free products (Qaim, 2009). However, several countries and regions have established a mandatory labeling system for GM foods, including the EU, Japan, and Russia.

The motivation underlying a mandatory approach is that consumers have a "right to know," which is different from the "need to know" approach in the context of risk communication. Moschini (2008) argues that the "right to know" approach is too open-ended and potentially unbounded because it can be invoked for virtually anything. The expected size of market segments is also a relevant question when deciding whether a voluntary or a mandatory labeling system should be preferred. When only a relatively small portion of consumers prefer GM-free foods, voluntary labeling will be much cheaper. A mandatory system becomes more attractive when the number of people who prefer GM-free foods increases. However, consumer preferences depend a lot on the information that consumers have available at a certain point in time (Huffman and McCluskey, 2014). When consumers are concerned about health risks of GMOs, they will prefer GM-free foods and a mandatory labeling system. On the other hand, a mandatory labeling system also fuels the notion that GM-foods are risky. A food that carries a mandatory GMO label must be perceived as risky by laypersons; why else would it be labeled in the same way as allergenic peanuts or cancer-provoking cigarettes (Davison, 2010)?

The EU has one of the strictest mandatory labeling regimes. Any food that is derived from a GM crop has to be labeled, regardless of whether or not the genetic modification is traceable in the end product. That is, vegetable oil derived from GM soy or sugar derived from GM sugar-beet would still have to be labeled, even though these products contain no protein to identify genetic structures. The food industry in Europe largely avoids such GMO-derived products. Strikingly, however, foods derived from animals that were fed with GM crops require no labeling in the EU. European farmers use a lot of soybean meal as a source of protein for their livestock. This soybean meal is primarily imported from North and South America, where almost all soybeans are genetically modified. The GM feedstuffs have to be labeled when used in the EU, but the meat, milk, and eggs eventually sold to consumers all remain unlabeled. Thus, hardly any GM-labeled products are found in European supermarket shelves. Non-food products derived from GM crops, such as textiles from Bt cotton, do not have to be labeled.

Many European consumers assume that food supply chains in the EU are largely free from GM products. This notion helps NGOs to depict GMOs as dangerous "Frankenfoods," which need to be rejected by all means. In reality, food supply chains in Europe are far from being GMO-free. A significant proportion of the feedstuffs used in European livestock production contain GM ingredients. Moreover, in food processing the use of GM microorganisms for the production of relevant enzymes has been common for many years. Foods derived from such processing techniques do not require labeling in the EU. Given the stalemate in the European debate it could be useful to change the system and require comprehensive labeling of all foods for which GMOs played some role along the chain, including livestock products fed with GM feedstuffs and foods processed with the help of GM microorganisms. With such a change, a large share of all food products sold in Europe would have to be labeled. Realizing that the consumption of GMO-derived products does not cause any harm for human health could contribute to gradually changing public attitudes. With this in mind it may not surprise that anti-biotech NGOs are lobbying against a more comprehensive and transparent labeling regime in the EU.

Several countries have introduced public or private voluntary labeling systems for GM-free products. Where neither mandatory nor voluntary labels for GMOs are available, consumers who resist GM foods can resort to organic food. The International Federation of Organic Agriculture Movements (IFOAM) and most national organic agriculture associations reject the use of GM crops as a matter of principle.

In the United States, food suppliers who wish are free to use positive or negative labels, yet without using the phrase "genetically modified." Instead, "genetically engineered" or "made through biotechnology" can be used. This is to avoid confusion among consumers because the FDA defines all breeding methods as forms of genetic modification (Huffman and McCluskey, 2014). The number of such labeled products in the United States is small. Until recently, US consumers were not much concerned about GM foods. This has changed recently in some consumer segments. Since 2012, several US states have allowed their residents to vote for or against new mandatory labeling laws for GMOs. In most states, the majority voted against mandatory labeling. In a few other states, such as Connecticut, Maine, and Vermont, voters were in favor of labeling. Vermont was the first state to pass a mandatory labeling law in 2014, taking effect in July 2016. However, this law was challenged in a federal court. That US consumers in some states have recently become more concerned about GM foods is not because of new evidence of risk, but due to the increasingly vociferous efforts of anti-GMO activists to convince the public that GMOs are bad. It remains unclear how the question of labeling and consumer and voter attitudes toward GMOs in the United States will evolve in the years to come.

Labeling involves market segregation and a system of identity preservation, which can be quite costly. The cost is negatively correlated with the threshold levels allowed for the adventitious presence of GM material. Again, these thresholds are not related to risks but are a political decision; very low thresholds can lead to prohibitive segregation costs. In the EU, the threshold for the adventitious presence of GM ingredients is 0.9 percent for GMOs approved for human consumption. Products with GM ingredients beyond that threshold have to be labeled. For GMOs not approved in the EU, the threshold is zero. Studies have shown that labeling in general and segregation costs in particular can influence the welfare effects of GM crops significantly (Qaim, 2009). Dissimilar approaches across countries can also lead to problems in international trade. Differences in the labeling of GMOs are among the issues discussed in the negotiations on the Transatlantic Trade and Investment

Partnership (TTIP), a planned free-trade agreement between the EU and the United States.

Labeling and segregation are also related to coexistence. The EU in particular has established rules to ensure the coexistence of GM crops with conventional and organic farming. These coexistence rules involve a number of technical and legal specifications, from minimum distance requirements for cultivation to liability and insurance measures (Beckmann et al., 2006). The high degree of complexity, uncertainty, and direct costs associated with these coexistence rules represent clear disincentives for EU farmers to adopt GM crops, even in those few cases where GM varieties are approved for cultivation (Devos et al., 2014).

Several developing countries are also considering the labeling of GM foods. A few have labeling laws already in place. China has mandatory labeling for certain products since 2002. India has recently passed a law that requires packaged foods to be labeled when they contain GM ingredients. However, only very few foods in these countries are actually labeled, although many do contain GM ingredients. China, for instance, imports large quantities of soybean from the United States. Enforcing reliable labeling systems for GMOs and effective market segregation is difficult even in emerging countries such as China and India. It will hardly be practicable in poorer countries of Africa and Asia because many foods are traded in informal rural markets. Effective monitoring could involve a prohibitive cost. The money in developing countries should rather be spent on more important areas of food policy. As only GM foods that were approved by the regulatory authorities are allowed to be grown and imported in a country, labeling is not about food safety, only about preferences.

Regulation and International Trade

Food and agricultural commodities are traded internationally, so that differences in regulatory approaches can also have important trade implications. Issues of international agricultural trade are generally governed by the World Trade Organization (WTO). However, when the first GM crops were commercially released in a few countries in the mid-1990s and the broader public debate about possible risks commenced, many felt that the WTO is not the appropriate institution to make trade rules for this transformative technology. As an alternative, the Cartagena Protocol on Biosafety was negotiated under the auspices of UNEP. As discussed earlier, the Cartagena Protocol was ratified by many countries, including major importers of agricultural commodities, but not by several of the major exporters, including the United States, Canada, and Argentina. The

Cartagena Protocol stipulates that the use of "living modified organisms" must be based on the precautionary principle. It allows countries to ban the import of GMOs on grounds of scientific uncertainty. The WTO, on the other hand, does not allow restricting trade based on the precautionary principle. Thus, there are competing international rules for governing the trade in GMOs and GMO-derived products (Kerr, 2014).

Countries that have signed the Cartagena Protocol only allow imports of those GMOs that were explicitly approved for importation and consumption by the national regulatory authorities. Similar to the approval procedure for the commercial release and cultivation of GM crops, the technology developer has to submit an application to the national authority of the importing country together with a regulatory dossier. Depending on the particular product, the focus of the assessment is on food and feed safety. In Europe, the EU Commission will also seek a scientific opinion from EFSA before initiating the further procedure with involvement of the EU member countries. Getting the Commission's approval for the import of a GMO or GMO-derived product is usually faster (and with higher probability of success) than getting approval for the cultivation of a GMO, but it can still take several years.

In practical terms, the developer of a GM crop must not only seek regulatory approval for cultivation in the countries where the crop shall be grown; he/she must also seek approval for import and use in all those countries where the crop may eventually be consumed. GM crop developers must also consider the sequence of application submissions, taking into account the requirements and expected durations of the approval process in each country (Kalaitzandonakes et al., 2014). As agricultural commodities are widely traded across countries and continents, these are increasingly complex and costly procedures for the innovator. Such procedures are easier to manage for multinational companies than for smaller firms or public sector research organizations. This is another mechanism through which GMO overregulation contributes to market concentration.

Delayed approval in an importing country can lead to serious trade disruptions. Between 1999 and 2003, the EU had implemented a de facto moratorium on GMOs. During this time period, any new approvals for both cultivation and import of GMOs were suspended. In the United States and other exporting countries, new GM technologies had been approved and adopted during this period, so that exports to Europe suffered. In Argentina, certain new GM technologies were not commercialized during this period in order not to jeopardize exports to Europe. The United States, Canada, and Argentina filed a complaint against the EU moratorium at the WTO. The WTO's Dispute Settlement Body ruled that the EU moratorium was indeed illegal under the Organization's rules. The EU Commission resumed considering GMO applications in late 2003, but approval processes remained much slower than most of the exporting countries had hoped for. Different speeds of GMO approvals in exporting and importing countries are often referred to as "asynchronous authorization" (Davsion, 2010).

Asynchronous authorization can become a particular problem for widely traded commodities such as soybean and maize. Segregation along the entire supply chain is required not only between GM and non-GM products but also between authorized and unauthorized GM products—classifications that often differ across various trading partners. As discussed earlier, there are now multiple GM events that are used in particular crops. For instance, if the United States exports soybeans to China, the EU, and other countries, and all these countries are asynchronous in terms of their import approvals for different GM events, then multiple supply chain channels have to be established. Segregation must prevent admixtures between approved and unapproved events at all stages—production, harvest, storage, processing, and shipment. This requires stringent management, cleaning, and testing procedures. The rigor with which segregation procedures have to be pursued depends mostly on the thresholds that importing countries have for unapproved GM events in their food and feed supply chains.

When exporting to countries with zero threshold policies, such as the EU, the segregation costs and the risk of encountering product failures are high (Kalaitzandonakes et al., 2014). This also includes potential liability issues for not complying with national regulations, even when only tiny traces of the safe but unauthorized event are found. For small trading companies these costs may be prohibitive. Thus, asynchronous authorization contributes to concentration not only in the agricultural biotech industry but also in the trading industry.

A concrete example may help to explain possible problems that may occur. In July 2009, the EU rejected a ship cargo of 180,000 tons of soybean meal coming from the United States because it was found to contain tiny traces of MON8807 maize. This GM maize event developed by Monsanto is a stack of different insect-resistance and herbicide-tolerance traits. It was declared safe for human consumption by FDA. Monsanto had also filed an application for import authorization of this event in the EU. EFSA had evaluated MON8807 and had not found any indications of potential toxicity or allergenicity. In spite of this positive scientific opinion by EFSA, the event had not been formally approved by the EU Commission. The traces of MON8807 found in the cargo were so small that they could be clearly categorized as inadvertent admixture. A case like this is referred to as "low-level botanical presence." Nevertheless, due to the zero threshold for unapproved events in the EU, the entire cargo was rejected (Davison, 2010). Anti-GMO activists like to depict cases like this as a scandal of food contamination, so there is also a large reputational risk involved for the trading companies as well as for the biotech companies owning the GM crop events.

The frequency of such incidents could increase with more and more GMO events being approved in exporting but not in importing countries. This could lead to large trade costs and disruptions. Under a zero threshold policy, trade of relevant commodities between countries with asynchronous authorization will likely cease. Alternatively, the prices that importing countries with zero threshold policies would have to pay could rise significantly. But obviously, especially when large importers such as the EU are involved, slow approval combined with zero thresholds can also reduce GM crop developments in the rest of the world. Developing countries in particular are hesitant to commercialize GM technologies when they are threatened to lose current or potential export markets. The non-approvals of insect-resistant rice in China and of virus-resistant papaya in Thailand are largely related to concerns about possible trade disruptions, coupled with targeted NGO campaigns. In African countries, trade relations with Europe also play an important role for national biotechnology policies (Paarlberg, 2014). Even in the United States, GM technologies in wheat were shelved in 2004 because of export market considerations.

There is also another aspect related to zero thresholds for unauthorized events that is of particular relevance for technologies developed by public sector institutions for poor people in developing countries. A case in point is Golden Rice that was developed to reduce vitamin A malnutrition in Asia (chapter 5). Golden Rice is not intended for use in Europe. However, as rice is traded internationally and small traces of inadvertent admixture cannot be ruled out with certainty, the safest option to avoid possible trade problems, which anti-GMO campaigners would love to scandalize, is to seek approval for import in the EU. This adds further barriers to humanitarian projects in terms of unnecessary costs and delays.

Regulatory asynchronicity will likely continue, but more sensible threshold policies in importing countries could help to avoid major trade disruptions in the future. Less politicized approval procedures would also help. Guidelines for abbreviated food safety reviews for GM crops that have been assessed and approved for commercialization in some countries have already been provided by the Codex Alimentarius (Kalaitzandonakes et al., 2014).

Intellectual Property Rights

Beyond biosafety and food safety rules, another type of regulation that applies to GM crops are IPRs. IPRs are also relevant for conventionally bred crops and other agricultural technologies, but their importance in the plant breeding industry has increased considerably with the advent of modern biotechnology. Plant breeding is different from other sectors in the sense that new plant breeding technologies can be reproduced more easily. A new chemical pesticide, for instance, cannot be reproduced by farmers themselves. Every time a farmer wants to use the chemical, he/she has to buy it, because once it has been sprayed it is gone. This enables the developer of the pesticide to recuperate the R&D investments through regular product sales.

The situation is quite different for new plant varieties. Farmers can easily reproduce seeds of most plant varieties by simply keeping some of the harvest and replanting it in the following season. This possibility of reproducing seeds is not changed through genetic engineering, that is, farmers can also reproduce GM varieties themselves. Without legal restrictions farmers can also share seeds with neighbors and even sell copies of the new variety in the informal seed market. This makes it difficult—or impossible—for developers of new plant varieties to recover their R&D investments. In such a situation, plant varieties are a typical public good and private companies have no incentive to invest in the development of new seed technologies. This will lead to lower than optimal innovation, unless the public sector invests sufficiently in crop improvement research. IPR protection is one way to deal with the problem of underinvestment by introducing legal use restrictions that substitute for the lacking genetic use restriction of most plant varieties.

Since the 1960s, breeders in many countries can seek plant variety protection (PVP) for their innovations. An internationally recognized framework for the protection of plant breeders' rights was established in 1961 under the International Union for the Protection of New Varieties of Plants (UPOV). While most developed countries are part of UPOV, many developing countries decided not to become part. Until the early 1990s, PVP laws maintained the right for farmers to reproduce seeds of protected varieties for their own use, but not for seed sales. Since then, several countries decided to extend the scope of PVP, requiring farmers to pay a license fee to breeders when using farm-saved seeds of protected varieties. One important characteristic of PVP laws is that they allow protected varieties to be freely used for further R&D purposes. That is, if breeder A uses a variety that is protected by breeder B for the development of a new variety with at least one added characteristic, breeder A can protect and sell this new variety without requiring a license from breeder B.

Large investments into biotechnology and the emergence of new tools of genome analysis and genetic engineering led to calls for a wider scope of IPR protection not only for plant breeding technologies, but in the broader biotech industry. In 1980, the US Court of Customs and Patent Appeals approved a patent on a genetically modified bacterium. This court ruling, which became known as the *Diamond vs. Chakrabarty* case, was a landmark decision, as it involved the first patent of a GMO (Just et al., 2006). Patent protection was later also extended to GM crops in the United States and other developed countries. Patent protection is stronger than PVP, as patented technologies cannot legally be used for follow-up research without obtaining a license from the patent holder. In addition, while PVP only applies to new varieties of plants, patents can also be granted for individual genes—such as Bt, promoter, and marker genes—and enabling technologies for plant transformation.

The option to obtain patent protection on biological inventions has spurred a tremendous amount of private sector plant biotech research since the early 1980s. Nowadays, more than 75 percent of all patents in agricultural biotechnology are held by the private sector, mostly by a handful of multinational corporations (Qaim, 2009; Graff et al., 2013). The proliferation of patents on plant biotech innovations has caused a lot of public concern and is one of the reasons for the opposition to GM crops. First, there are widespread ethical concerns with patenting life and genetic materials that exist in nature. Second, there are social concerns because it is feared that patents lead to corporate control of the food chain, seed monopolies, and exploitation of farmers.

Whether patents on life are right or wrong is a question for which a global consensus may be hard to reach. This cannot be discussed for plant biotechnology in isolation because the same issue also applies to various other sectors where GMOs and biological inventions more generally play an important role, such as the medical and pharmaceutical industries, food processing, material science, environmental science, and bioenergy. Biotechnology is one of the key technologies for the twenty-first century with large potential to contribute to greening economies. Hence some form of IPR protection will be required to provide private sector incentives for sufficient R&D investments.

Concerning the social concerns, many of the arguments build on false assumptions, as will be discussed below. But even if one draws the conclusion that patents on GM crops have negative social consequences, this would explain reservations against patents, not the opposition against GM crops in general. Without patents there would be lower private sector

R&D investments, but public research organizations could develop GM crop technologies nevertheless.

Will patent protection of GM crop varieties lead to the exploitation of farmers, as is often assumed? For exploitation to occur, two conditions must hold. First, patented seeds are sold at a very high price that leaves no benefit for farmers. Second, farmers are forced to buy the patented seeds, that is, they have no alternative. Both conditions are unlikely to hold. Patent protection provides the innovator with a temporary monopoly for the particular technology. Thus, a seed variety carrying this technology will be sold at a premium. But the seed price cannot be higher than the average benefit that farmers would derive from adopting this variety; otherwise farmers would decide to adopt alternative varieties, including conventional ones. Hence, the monopoly is a restricted monopoly (Basu and Qaim, 2007). But what if the seed price is low in the beginning, to entice farmers to adopt, and then later the price is raised? Farmers would simply disadopt the GM variety and use conventional seeds again. The choice of which seeds to use is made every season, hence the decision to adopt a GM variety in one season is not irreversible if economic conditions change in subsequent years.

Will alternative varieties always be available? As seen in chapter 4, there are examples of countries where the large majority of varieties for a particular crop are now genetically modified, and GM adoption rates are close to 100 percent. However, such high adoption rates occur in situations where GM seeds are sold by many seed companies at affordable prices and benefits for farmers are sizeable, not in monopoly situations with excessive GM seed prices. These considerations and predictions are confirmed by a large number of empirical studies carried out under different IPR regimes (Qaim and de Janvry, 2003; Qaim, 2005; Qaim and Traxler, 2005; Just et al., 2006; Smyth et al., 2014; Qiao, 2015; Krishna et al., 2015).

A hypothetical scenario from the market for medical drugs may help to further clarify some important differences. Consider a pharmaceutical company that has obtained a patent for a life-saving drug. Let us further assume that for the deadly disease that this drug can treat no other effective treatment is available. In that case, people suffering from this disease have little choice but to purchase the drug, even if the price charged by the patent-holding company is excessive. This could be judged as a case of customer exploitation. On the other hand, poor people who cannot afford to buy the drug would be doomed to die. However, a scenario like this cannot be transferred to the markets for agricultural seeds because there is no new seed variety that would make an immediate difference between life and death. Hence, excessive prices charged for a new seed variety would lead to low adoption rates, but not to farmer exploitation, as long as different sources of seed supply are available to farmers. Monopolization of entire seed markets—beyond individual varieties and technologies—has to be avoided, but this holds true with and without patenting, and with and without GMOs.

Another common misconception relates to the geographical scope of patents. Patent law is national law, so that a patent granted in one country does not hold automatically also in other countries. Many GM crop technologies are patented in the United States, Canada, and other rich countries, but often not in developing countries, either because the innovator did not apply for a patent there or because a patent was not granted by the national authorities. As discussed earlier, Roundup Ready soybeans are patented in the United States, but not in Argentina and several other countries in South America. Likewise, Bt cotton technology is patented in the United States, but not in India and China. This is also one of the reasons why in most developing countries prices for GM seeds are lower and financial benefits higher than in developed countries (Klümper and Qaim, 2014).

Companies often lobby for strengthening IPR protection in developing countries. This could contribute to higher private R&D investments and possibly more location-specific innovation in countries where the overall investment climate is favorable. However, seed prices would likely increase. The appropriate level of IPR protection varies by country. Especially in the least developed countries, the drawbacks of strong IPR protection will likely outweigh the advantages. Hence, poor countries should be careful with establishing too strong IPR laws. From a social perspective there is nothing wrong with companies making more money with their patented technologies in rich countries, while the same technologies can be used by smallholder farmers in poor countries at lower costs.

In addition to patenting new GM crop technologies, innovators can also obtain patents for genes that were isolated and characterized and for enabling technologies such as plant transformation or tissue culture techniques. This is of less immediate relevance for farmers, but has important implications for the freedom-to-operate within the biotech industry itself. Since the development of a single GM crop may require the use of dozens of patented genes and enabling technologies, researchers—or their legal representatives—have to negotiate licenses with multiple patent holders, involving high transaction costs (Qaim et al., 2000). In that sense, the freedom-to-operate problem likely contributes to further industry concentration. In some cases, it may be easier for a large company to buy up smaller companies with interesting patents, rather than engaging in complex licensing negotiations. Large companies also have an advantage when it comes to cross-licensing, not only because of their broad patent portfolios but also because of their extensive legal experience.

To reduce transaction costs and support public sector organizations in patent licensing, IPR clearinghouse mechanisms have been developed. One example is the Public Intellectual Property Resource for Agriculture (PIPRA), which collaborates with over 50 universities and research centers to reduce IPR barriers and facilitate technology transfer, with a particular focus on benefiting developing countries. Such public sector initiatives are important, as there are certain research and technology areas that will not be addressed by private companies because of the limited size of potential markets or other types of constraints. Examples are technologies especially designed for poor farmers and consumers. In such areas, public research is needed. Moreover, public-private partnerships (PPPs) are useful to harness the comparative strengths of both sectors.

There are many interesting examples of PPPs to develop GM crops for humanitarian purposes. These include the development of drought-tolerant maize, nitrogen use efficient and salt-tolerant rice, and biofortified sorghum for Africa, just to name a few (chapter 5). Companies are often willing to bring in their patented technologies royalty-free to such PPPs for humanitarian purposes. One of the first PPPs that resulted in a commercialized GM product is Bt eggplant, which was developed by the Indian seed company Mahyco and transferred to the public sector in Bangladesh with support from USAID (James, 2014).

Regulation and the Structure of the Biotech Industry

The global market for GM crop traits is quite concentrated with a few multinational companies dominating development and commercialization. The market leader is Monsanto accounting for the largest number of GM crop events, followed by Bayer CropScience, Dow AgroSciences, Syngenta, and Pioneer/DuPont. Most of these companies have a background in the chemical pesticide industry. In comparison, the markets for seeds are much less concentrated because local companies and public breeding stations still play a more dominant role for germplasm development that needs to be well adapted to particular agroecological conditions. Even for GM seeds, the number of companies developing and selling the GM varieties is much larger than the number of companies developing the GM traits, as the backcrossing of approved GM traits into locally adapted germplasm is a relatively straightforward procedure. If the GM trait is patented in the relevant jurisdiction, seed companies that wish to backcross the technology into their own varieties have to obtain a license from the patent holder. If there is no valid patent, a license is not required.

The separation of trait and germplasm development is a phenomenon that is directly connected to GM technology. Before the advent of modern biotechnology, germplasm development and trait development were inextricably linked and carried out by public or private sector plant breeding organizations. The breeding output was a new variety that displayed certain desirable characteristics for a particular region. With genetic engineering this has changed. GM traits cannot only be introgressed into a large number of varieties, they can also be transferred to other crop species. Thus, GM trait development is a business that is global in scope, while germplasm development often remains a local business. Unsurprisingly, these technological developments also had important effects on market structures.

In the 1980s when the research on GM crops started, plant breeding and molecular biology were two completely different disciplines. At that time, most breeding companies and institutions did not have much capacity in molecular biology, so that it was not the plant breeding sector that started GM trait development. Some of the pioneering tools for plant biotechnology were developed in specialized public sector laboratories. Several of these academic innovations led to startup companies. At the same time, the agrochemical industry started with related biotech research because it was realized that GM traits could become an important new tool for plant protection in the future.

The agrochemical industry had always been dominated by a few large companies because the development of chemical pesticides is a very complex and costly business, which also requires substantial research on potential environmental and health risks. The human and financial capacity to carry out sophisticated lab research and compile regulatory dossiers for biosafety and food safety authorities proved to be a great advantage to successfully develop and commercialize GM traits. Thus, the agrochemical multinationals also became plant biotech multinationals. Some of the smaller biotech startups with limited financial capacity were later also acquired by these multinationals (Heisey and Schimmelpfennig, 2006).

However, the agrochemical companies had no experience in the seed business. They also did not have access to good germplasm. Thus, they needed a strategy of how to get their new GM traits successfully out to farmers. One option was to license the GM traits to existing seed companies. The other option was to acquire seed companies that owned important germplasm for trait introgression and hence market GM varieties directly. Both options were pursued by agrochemical and biotech companies, often in combination. A large number of mergers and acquisitions occurred since the 1990s. Especially in North and South America, many small and medium sized seed companies, and also a few larger ones, were purchased by biotech multinationals. Several seed companies also merged horizontally, to be in a better strategic position. A few seed companies also started their own biotech investments (Fernandez-Cornejo, 2004).

A little bit of history of two of the leading GM trait and seed companies—namely Monsanto and Pioneer/DuPont—may help to better understand the enormous dynamics in industry structure. Monsanto was founded in 1901 as a US chemical company. It focused initially on the production of artificial sweeteners and flavors for the food industry. The Company produced and marketed its first agricultural pesticide in the 1940s. In the 1960s and 1970s, Monsanto had developed several herbicides, most of which have remained quite successful in the market since then. Already in the mid-1970s, the company established a plant biotechnology laboratory. In this lab, in the early 1980s the first GM crop worldwide was produced using the *Agrobacterium* transformation process. Monsanto was also the first organization in the United States to carry out a field trial with GM crops in 1987, and to introduce a Bt gene into cotton coding for insect resistance in 1991. Herbicide-tolerance traits were also developed.

Initially, Monsanto licensed its insect-resistance and herbicide-tolerance traits to existing seed companies that started selling GM varieties of cotton, maize, and soybean in the mid-1990s. However, from 1997 onward Monsanto started buying seed companies itself in the United States and elsewhere, including large companies such as Holden, DeKalb, Cargill, Delta & Pineland, and others. In 2000, Monsanto merged with the pharmaceutical company Pharmacia & Upjohn; two years later the agricultural part was spun off again, retaining the Monsanto name. In the mid-2000s, Monsanto purchased Seminis, the world's largest vegetable seed company, and also invested in several seed businesses in South America (Huffman, 2011). In 2015, Monsanto started negotiations to acquire Syngenta.

A very different example is that of Pioneer/DuPont, the world's largest seed company. Pioneer was established in 1926 as the Hi-Bred Corn Company. The Company, which was renamed Pioneer Hi-Bred Corn Company in 1935, was instrumental in developing hybrid maize. Over the following decades, Pioneer developed a very large germplasm inventory for maize. In the 1970s, the company also expanded its business into a few other crops, including soybean and alfalfa (Fernandez-Cornejo, 2004). Pioneer started to invest in biotech research in the late 1980s. In the early 1990s, Pioneer purchased licenses to use Monsanto's Roundup Ready and European corn borer resistance technologies in its maize germplasm.

In the mid-1990s, Pioneer acquired parts of Mycogen, another seed and biotech company, with the intention to gain access to additional Bt genes and traits. In the late 1990s, Pioneer was itself purchased by the chemical company DuPont (Huffman, 2011). Over the last 15 years, Pioneer/DuPont has invested significantly to develop its own GM traits.

In Europe, agrochemical and biotech companies, including Syngenta, Bayer CropScience, and others, were also involved in various mergers, acquisitions, and spinoffs. It is quite obvious that GM crops are a transformative technology that has caused significant structural changes in various industries. Costly, lengthy, and in many cases unpredictable biosafety and food safety regulations and the proliferation of patents have contributed to further industry concentration. How exactly the biotech industry would look like with simplified and more sensible safety regulations is unclear. What is clear, however, is that with the current politicized regulatory approaches observed in Europe, only very large companies with substantial financial capacity can successfully commercialize GMOs.

There is some evidence that large agrochemical and biotech companies initially welcomed or at least tolerated the establishment of very strict safety standards for GMOs and rising regulatory hurdles, even when it was clear that this contradicted the scientific evidence (Taverne, 2007; Ammann, 2014). On the one hand, it was believed that tight regulation would raise public confidence in the safety of this technology. In reality, it had the opposite effect. "If GMOs are regulated so strongly, then they must be really dangerous," goes the public belief. On the other hand, some of the multinationals hoped that complex and costly regulatory procedures would help to keep smaller competitors out of the business of commercializing GMOs. This expectation has materialized indeed.

The relatively high level of concentration in crop biotechnology and related industries is an area of concern. Market power should be avoided, as sustainable innovation requires fair competition and affordable access to technologies. However, using this as justification to ban GMOs seems odd. This would be similar to banning computers because Microsoft has some market power in the software industry, or to ban the Internet because of the size of Google. Much better responses to foster competition are sensible antitrust policies and simplified, product-based safety regulations. Some reform in IPR laws may also be necessary.

CHAPTER 7

THE COMPLEX PUBLIC DEBATE

Public attitudes toward GM crops are predominantly negative. This is especially true in Europe, but European perceptions have also spread to other parts of the world. Many people do not only believe but are strongly convinced that GMOs do not bring any benefits for farmers and consumers. Instead, GMOs are seen as a technology that is dangerous for human health and the environment and that contributes to monopolies and corporate control of the food chain, thus causing new dependencies and other social problems. The empirical evidence discussed in previous chapters clearly shows that this notion is wrong. Commercialized GM crops have already produced significant benefits for farmers, consumers, and the environment, and they have an unblemished safety record. Thirty years of risk research also suggest that GM crops are not inherently more risky than conventionally bred crops. If used inappropriately, GM crops can create certain problems, but the same holds true for any other technology. How comes then that public perceptions differ so vastly from the scientific evidence and that this rift has actually further increased over time? The answer is that a huge protest industry against GMOs has emerged since the 1990s. This protest industry strongly influences public opinions and policymaking.

The anti-GMO protest industry is led by a few international NGOs such as Greenpeace and Friends of the Earth. These and other international NGOs also influence the agenda of local NGOs in developing countries, including many grassroots organizations. Anti-biotech campaigning NGOs create fears by inventing risks and deliberately misinterpreting scientific results. For them it does not matter whether dozens of academic studies have proven their claims wrong; by simply repeating the same claims again and again, if necessary underpinned with dubious evidence, they gradually become public truths. NGOs are often well financed for their campaigns, very skillful in their public communication, and extremely successful in their outreach and influence. The mass media happily reports about NGO stories, because fears and horror scenarios sell much better than boring reports about safe and beneficial new crop technologies. Analysis shows that repeating familiar stories, perpetuating stereotypes, and playing on public anxieties can help to raise media revenues (Curtis et al., 2008). Pictures of NGO stunts and activists with white protection suits and respiratory masks vandalizing GMO trials are powerful tools to fuel the notion of extreme danger. Buzzwords such as "Frankenfoods," "terminator technology," and "superweeds" also serve the purpose of creating and maintaining fears. NGO narratives about adverse GMO impacts are also picked up by food writers, such as Michael Pollan and Marion Nestle, in their widely read books and Internet blogs.

It helps the protest industry that GM crops are primarily developed and commercialized by private sector multinationals. There is widespread public distrust against private companies, and the larger a company, the larger is the level of distrust. Statements about the safety and usefulness of GM crops made by company representatives are regularly dismissed as industry propaganda. Such company statements are not a credible counterweight to NGO claims in the public debate. To many, statements by Monsanto, Syngenta, or Bayer CropScience about sustainable farming and reductions in agrochemicals through GMOs are automatically perceived as nonsense, given these companies' histories in selling chemical pesticides. That the first GM trait commercialized and widely adopted was herbicide tolerance was not at all helpful to increase public acceptance. Herbicide tolerance may be a useful technology from an agronomic perspective, but it is not an easy sell to the wider public. GM crops developed to withstand broad-spectrum herbicides produced and marketed by the same companies, such that more of these herbicides can be sold, just fit the public expectation of environmentally and socially unscrupulous pesticide companies too well.

NGOs have successfully framed GMOs as something inherently evil, and this evil is epitomized by Monsanto. This becomes very obvious by NGO-initiated movements named "Operation Cremate Monsanto" or "March against Monsanto" (Herring, 2007; March against Monsanto, 2015), which are quite militant in their activities yet enticing a sizeable community of followers. A dualistic mindset of "profit versus people" or "profit versus the environment" is a useful approach to simplify a complex topic and make it easy for uninformed persons to quickly form an opinion (Aerni, 2014). Fighting against the big evil has to be right, so that it becomes unnecessary to form a differentiated picture of a theme that is too complex to understand for the casual observer. Against this background, the option that GM crops could possibly contribute to sustainable agricultural development is considered completely absurd. In the framing of the protest industry, GMOs serve as a proxy for all the things that contribute to unsustainable farming.

The public notion of GMOs being inherently evil has become so widespread and entrenched that people with different opinions and arguments are often intimidated by the emotional confrontation. This even holds for researchers to some extent. Some publish their papers in academic journals, but otherwise keep silent, in order not to be confronted with harsh and emotional public opposition. Academic journals are not read by the wider public, so there is no danger in publishing there. But newspaper articles, public speeches, and radio or TV interviews are followed widely. I have myself experienced several cases of concerned citizens, who—after reading about a new study from my research group in a local newspaper—called my home phone number and left messages urging me to stop this type of research for the sake of our children. Of course, a few researchers do communicate publicly, but their statements alone are not enough to change deeply rooted prejudices.

Yet it would be short-sighted to believe that the GMO debate is merely a fight between NGO activists and the biotech industry, because there are several groups who benefit tremendously from the negative public sentiments against this technology (Apel, 2010). The organic food industry is a clear winner of the GMO protest industry. While organic farming could actually become more sustainable from assimilating pest- and disease-resistant GM crops (Ronald and Adamchak, 2008), the organic industry had soon realized that the economic benefits of banning GMOs from organically certified food production would be far greater. Especially in the United States, where no mandatory labeling for GM foods exists, consumers who want to avoid GMOs have switched to organic. Also other large food companies and retailers have launched GMO-free labels, thus benefiting from the anti-biotech sentiments of consumers.

Another group that benefits from the opposition to GMOs is the chemical pesticide industry, because widespread GM crop adoption tends to reduce farmers' demand for chemical pesticides. While a handful of large agrochemical companies have become biotech multinationals, so that lower revenues from pesticides can be compensated through higher revenues from seed sales, many smaller pesticide companies suffer a lot from increasing GM crop adoption. In India in particular, chemical companies have supported the opposition against Bt cotton. Also elsewhere, chemical companies lobby governments to restrict the use of GM crops. Finally, politicians can also benefit from widespread public fears. Decisions to ban GMOs or to seriously restrict the use of this technology are quite popular and cheap to implement, when the foregone benefits from not using the technology are not taken into account. Graff et al. (2009) have described this coalition of NGOs, organic and conventional food industries, agrochemical manufacturers, and certain politicians, who all have incentives to negatively characterize GMOs, as a "strange bedfellows constellation of concentrated economic interests."

In the following sections, I review some of the persistent NGO narratives about GMOs and how they have shaped public opinions globally in spite of strong scientific evidence to the contrary. I refer to NGOs frequently and mean non-governmental organizations that oppose GMOs. There is a huge number of NGOs, probably more than one hundred thousand globally. I recognize that not all of them have negative attitudes toward GMOs, but I use the abridged terminology to keep the text readable.

Narratives about Health Impacts of GMOs

NGOs like to spread messages that GM crops are linked to allergies, cancer, birth defects, sterility, and many other health calamities. People are increasingly concerned about the safety of their food, so that many try to avoid GMOs if they are continuously confronted with such messages. While consumer concerns about food safety are completely understandable, fears are often based on myths and wrong risk perceptions, a phenomenon that is observed well beyond GMO foods. At least in Europe, many consumers believe that modern agriculture has made their food increasingly unsafe. Apart from GMO ingredients, chemical pesticide residues are subjectively ranked by consumers among the biggest health risks associated with food. What most people do not know is that the foods produced and consumed today are much safer than the foods available at any time in human history. Especially in developed countries, where strict food standards are enforced, the health risks of pesticide residues and other human-made inputs are negligible and much lower than the risks associated with fungal or bacterial contamination.

As explained earlier, GM crops are tested much more comprehensively than any other technology, so that they are as safe—or even safer—than conventionally produced foods. Nonetheless, the direct changes in the genetic makeup of plants create anxieties of unpredictable long-term effects among those unfamiliar with the science, a phenomenon which is exploited by anti-GMO activists. What these activists deliberately fail to highlight is that the changes in the genetic makeup of plants through conventional breeding methods are often much more profound and less precise than those made through genetic engineering.

Occasional research studies that seem to provide evidence of negative health effects of GMOs help in nurturing the NGO narratives. Such studies are widely quoted and promoted on NGO websites. A few of these studies were published in respected academic journals, but were later on heavily criticized by independent scientific organizations. One example of such a controversial study was a paper by Stanley Ewen and Arpad Pusztai that was published in the medical journal *The Lancet* in 1999. The study had looked at health effects of a GM potato variety that was not commercialized and not intended for human consumption in a feeding experiment with rats. Before publication of the paper, Pusztai reported about effects on the rat's immune system in the media. However, these effects on the immune system were not mentioned in the actual paper. In the paper itself, differences in the rats' gut epithelium were reported, which the authors interpreted as a result of the potato transformation. Several scientific rebuttals on the study were published. The British Royal Society reviewed Pusztai's data and concluded that the study was based on flawed design, execution, and analysis (Loder, 1999). Up till today, the Pusztai results are frequently used by anti-biotech campaigners to underline the alleged health risks of GMOs.

More recently, a study led by Gilles-Eric Séralini claimed negative health effects of GM maize. This study was published in the journal *Food and Chemical Toxicology* in 2012. For the study, a small sample of rats was fed commercialized herbicide-tolerant maize over a two-year period. The authors reported that the rats developed more tumors and died earlier than the controls and concluded that GM maize is toxic. However, the Sprague-Dawley strain of rats chosen for the experiment are known to suffer from high and early rates of cancer development, so that they are not suitable for long-term feeding studies. The Séralini publication was heavily criticized for its inadequate experimental design, the small sample size, and the highly misleading conclusions (Arjó et al., 2013). Also official regulatory bodies, such as EFSA, Germany's Federal Institute for Risk Assessment, and similar national authorities in Belgium, France, Denmark, and other countries, disapproved the publication for providing inadequate data to support the authors' conclusions. In 2013, the journal decided to formally retract the paper (Casassus, 2013).

This Séralini study is noteworthy not only because of the poor science but also for the way it was communicated and handled by the media. Before the paper was initially published in 2012, Séralini approached media journalists and shared the manuscript together with pictures of tumor-ridden rats. Journalists agreed to sign a confidentiality agreement prohibiting them to contact other scientists for getting additional opinions prior to the paper's publication. This is a very unusual procedure.

The study authors also asked the journal to delay publication while they organized a press conference and prepared a video titled "Are we all guinea pigs?" that was part of the communication campaign (Arjó et al., 2013). On the day of publication, reports about the study together with the frightening pictures and video sequences were widely covered by the mass media, including in prime-time TV news. Due to the confidentiality agreements, any critical voices were not available on that day. The criticism that followed and the paper's retraction was discussed among experts but hardly covered by the large media houses. What remains for the wider public is the notion that GM maize was shown to cause cancer, at least in rats.

There is also the widespread public perception that glyphosate, the broad-spectrum herbicide that is used in combination with most of the so far commercialized herbicide-tolerant GM crops, is highly toxic for human health and the environment. This alleged glyphosate toxicity is used by NGOs as an argument against the deployment of GM crops. Related to this, rumors that heavy sprays of glyphosate have contributed to higher rates of birth defects and child cancer in soybean-growing regions of Argentina were spread and picked up by the media (Antoniou et al., 2010). A small epidemiological study carried out in Paraguay was used in support of this claim. However, the study from Paraguay looked at chemical pesticides in general and did not even mention glyphosate (Palma, 2011).

Glyphosate has been used already for more than 40 years and is known to be of low toxicity. In the World Health Organization (WHO) classification, glyphosate is included in the lowest toxicity class for "practically non-toxic" pesticides. In contradiction to a large number of articles that failed to establish negative health effects, one study published in 2010 by Alejandra Paganelli and coauthors reported malformations in frogs and chicken through glyphosate exposure. This study created widespread fears and led to calls for an immediate ban of glyphosate (Antoniou et al., 2010). But Paganelli and her colleagues had injected high doses of glyphosate directly into chicken and frog embryos in the lab, which is a route of exposure that would never happen under practical conditions. Hence, the study was found not suitable or relevant for risk assessment for humans and wildlife (Palma, 2011).

Recently, the debate about the health risks of glyphosate was fueled again through an evaluation made by the International Agency for Research on Cancer (IARC), a panel of the WHO. In March 2015, IARC published the summary of a study in which glyphosate was classified as "probably carcinogenic to humans" (IARC, 2015). Again, this publication led to immediate calls for a complete ban of glyphosate by several environmental groups and anti-biotech activists. However, hasty policy reactions may be inappropriate. IARC's report contradicts pesticide-regulating bodies—such as the US Environmental Protection Agency (EPA), the European Food Safety Authority (EFSA), the German Federal Institute for Risk Assessment (BfR), and the WHO/FAO Joint Meeting on Pesticide Residues (JMPR)—that all conclude that glyphosate does not cause cancer in humans and animals.

Many international toxicologists have criticized the IARC study's approach and methodology (Academics Review, 2015). First, the assessment seems to build on a review of only a small number of selected studies, while many other available studies about the health effects of glyphosate were not considered. Second, IARC's classification is based on what toxicologists call a hazard assessment, that is, the potential impact was evaluated without considering the dose at which humans are typically exposed to the chemical. A hazard assessment is different from a risk assessment that takes into account typical doses of exposure. Many substances are known to cause cancer at higher doses, which is why rules and recommendations for their use and exposure were established. Examples include salted fish, many paints and hair dyes, diesel exhaust, wood dust and leather dust, among many others. IARC (2015) acknowledges that the glyphosate dose that the general population is typically exposed to is low. Even the dose that farmers who spray glyphosate regularly are typically exposed to is well below critical levels.

On behalf of EFSA, BfR recently re-evaluated the toxicity of glyphosate: in January 2015, it completed a report reviewing over one thousand toxicological studies and concluding that there is no new evidence that glyphosate could cause cancer or birth defects in humans (BfR, 2015). Glyphosate is one of the most comprehensively evaluated chemical herbicides. While all pesticides may have negative effects on human health and the environment when applied at high doses, glyphosate is less toxic than most other pesticides that are widely used in agricultural production, including many plant protection products used in organic farming.

It is quite obvious that many of the recent claims about the toxic effects of glyphosate were made by anti-biotech campaigners with the intention to fuel public concerns about GMOs. Now that new GM soybean and maize varieties with tolerance to other herbicides are being commercialized, similar claims are launched for these other herbicides as well. A case in point is the herbicide 2,4-D, which has also been on the market for several decades. While NGOs depict 2,4-D as dangerous, EPA has evaluated this herbicide repeatedly and has consistently found it to be safe for human health and the environment.

In specific situations, other narratives about negative health effects of GM crops were constructed by anti-biotech NGOs. I often travel to Kenya, including to semi-arid parts of the country that are regularly affected by drought. In drought years, food aid is often distributed. This food aid also contains GM maize coming from the United States. It has been observed that in drought years rates of food poisoning increase, which local anti-biotech activists like to attribute to the consumption of GM maize. However, in reality the poison incidents stem from the locally harvested maize. In drought years, the little maize that grows is harvested very early and then stored in the house for fears of theft. These conditions promote the proliferation of aflatoxin and other mycotoxins, which are known to cause cancer and other health problems. The activists' claim of poisonous GM maize also ignores the fact that the same maize is eaten by US consumers on a regular basis.

Another popular narrative relates to sheep that allegedly died in Andhra Pradesh, a southern state in India, after grazing in GM cotton fields. Local and international NGOs, as well as Organic Consumers Associations in the United States, widely spread this story claiming that GMOs are a danger for animals and humans (Herring, 2010). What they left unexplained is why sheep outside of Andhra Pradesh, where the same GM cotton was grown, were not at all affected by this mysterious disease. While it is easy to unmask such NGO hoaxes from an international perspective, it is also easy to see how such narratives can undermine local people's trust in the safety of GM technology.

Narratives about Environmental Impacts of GMOs

Narratives about negative environmental effects of GM crops relate primarily to gene flow and "genetic contamination" of natural ecosystems and effects of transgenes on non-target organisms. These risks were already discussed in chapter 3. It is easy and popular to invent horror scenarios of superweeds that destroy natural biodiversity or GM crops that decimate populations of honeybees and other beneficial insects. Such scenarios may appear plausible for at least two reasons. First, the science of gene flow is complex and hard to understand for the layperson, whereas a picture of a giant maize plant that dominates everything else, such as often portrayed in NGO stunts, is simple and memorable. Second, new agricultural technologies in the past have often contributed to environmental problems, so why should it be different with GM crops? In spite of their apparent plausibility, such horror scenarios are not supported by any scientific evidence. No single case is known where outcrossing transgenes from herbicide-tolerant or insect-resistant GM crops have widely spread and caused biodiversity loss in natural ecosystems. This may possibly be different for crops with other traits. As discussed, the fitness advantage of new traits should be assessed case by case whenever wild relatives of the domesticated crops exist in the natural environment. But this holds true for both GM and conventional crops with new traits, as both are associated with the same types of risks (Raven, 2010; House of Commons, 2015).

An early study that sparked a lot of debate about gene flow and its consequences was a paper published in *Nature* in 2001 by David Quist and Ignacio Chapela. In this paper, the authors found that native maize in southern Mexico contained transgenes from GM maize, even though GM maize had officially been banned from cultivation in Mexico since 1998. The paper authors also reported that the transgenes were unstable in the genome and behaved in unpredictable ways. Especially this second result was heavily criticized by several other scientists who argued that the analytical method used was prone to artifacts and misinterpretation. Based on this criticism the journal officially announced in 2002 that the article should not have been published (Graham, 2002). But anti-biotech groups still used the Quist and Chapela study to create fears of the loss of the diversity of landraces in southern Mexico, which is the center of biodiversity for maize. This debate heavily influenced biosafety policies in Mexico and many other countries.

Gene flow between the maize that farmers grow and wild relatives occurs, with and without GM crops. As GM maize was officially cultivated in Mexico before 1998, finding transgenes in native landraces should not surprise. The more relevant question is about the consequences of such gene flow. Farmers in southern Mexico have grown improved varieties and hybrids of maize for many decades. Genes from these modern varieties are found in local landraces, as one would expect, but these landraces have not perished nor have they lost their identity. There is no reason to expect that the flow of transgenes from GM crops would have any other effect on the genetic diversity of landraces than the flow of other genes from conventionally bred modern varieties (Parrott, 2010; Raven, 2010).

The widespread myth of gene contamination builds on the assumption that landraces and natural biodiversity are static entities, but they are not. Gene flow between different landraces and also between wild and domesticated crops has always occurred and is part of normal evolutionary processes. This does not mean that modern agriculture did not significantly reduce biodiversity and ecological functions. It did and does through the abandonment of hedgerows and natural landscape elements, the narrowing of crop rotations and varietal diversity, and the heavy use of agrochemicals. But these developments started long before GM crops were introduced. As discussed in previous chapters, if properly used GM crops can actually help to reduce some of these environmental issues.

Concerns about the effects of GMOs on non-target organisms primarily relate to Bt crops. Bt proteins are much more selective than chemical insecticides, which is also one reason why Bt is sprayed as a biological pesticide to control insect pests in organic agriculture. Nonetheless, harm to non-target insects of the same or related orders may theoretically occur. An early study published in 1999 in *Nature* by John Losey and colleagues received widespread attention. The authors had shown in a lab experiment that larvae of the monarch butterfly feeding on milkweed leaves dusted with Bt maize pollen grew more slowly and suffered higher mortality than the controls. The lab conditions were not mimicking practical field conditions, but many still hastily concluded that Bt maize would harm the monarch butterfly, a charismatic insect. Comprehensive follow-up research showed that the impact of Bt maize on monarch butterfly populations in the environment is negligible (Sears et al., 2001).

The debate about impacts of Bt crops on other non-target insects— such as honeybees, lacewings, and ladybirds—continues. Numerous studies have been carried out on this issue, which is now probably one of the most analyzed environmental effects of any agricultural technology. Similar to the Losey article, a few studies that were carried out under artificial lab conditions showed some negative impacts, while these effects were not confirmed under practical field conditions. A meta-analysis of field studies showed that non-target insects are more abundant in Bt fields than in fields with conventional crops that were sprayed with chemical insecticides (Marvier et al., 2007). The same study also showed that certain non-target species were somewhat less abundant in Bt fields when compared to conventional crops without any chemical pesticides. However, without pest control yields are also much lower in most situations, so that this comparison is of little practical relevance.

Finally, an argument often used by NGOs to prove that GMOs are not sustainable is the possible resistance development in pest and weed populations. As discussed in previous chapters, resistance development can indeed occur when good agricultural practices are not followed and refuge strategies are not properly implemented. However, using this as a knockout argument against GM crops is inconsistent, because resistance development is an issue in almost all pest control strategies, regardless of whether genetic, chemical, or biological tools are used. Organic farmers in particular love to use genetic host plant resistance, as long as this was not achieved through genetic engineering. But conventionally bred crops with inbuilt pest or disease resistance are also not immune to resistance development.

Narratives about Social Impacts of GMOs

There are many false claims about negative social impacts of GMOs made by anti-biotech activists. Many of these claims stem from India or are related to Bt cotton and smallholder farmers in India. One of the most vocal figures in this debate is Vandana Shiva, who founded the Indian NGO Navdanya and has fought against the Green Revolution, trade liberalization, and the influence of foreign companies in India for several decades. Shiva has emerged as an anti-GMO celebrity at the global level and widely tours around the world to give speeches about the evils of modern agriculture in general and GMOs in particular.

The claim of thousands of Indian farmers committing suicide after adopting Bt cotton goes back to Vandana Shiva and is widely echoed by NGOs around the world (Shiva et al., 2011). I was shocked to see that these claims were also recently broadcasted through large public sector TV and radio channels in Germany. It was discussed in chapter 4 that the alleged link between suicides and Bt cotton adoption is pure propaganda and that millions of farmers in India benefit significantly from GM technology adoption. Several studies have dismissed Shiva's demagogic claim (Gruère and Sengupta, 2011; Gilbert, 2013; Qaim, 2014). But this evidence does not matter to her. In interviews and speeches she maintains: "270,000 Indian farmers have committed suicide since Monsanto entered the Indian seed market, it's genocide." Interesting to note is that Shiva wrote a book titled *Seeds of Suicide* already in 2000, before GM cotton had been introduced and before Monsanto had entered the Indian seed market. In that book, conventional hybrid seeds and globalization were depicted as the root causes of farmer suicides in India.

But why does the narrative of failing GM seeds and ruined farmers persist is spite of the large scientific evidence to the contrary? And even if one does not know or believe the scientific evidence about the benefits for farmers, how could one explain that Bt cotton adoption rates in India have increased year after year, rapidly reaching a level of 95 percent? Vandana Shiva and other anti-biotech activists know that their line of argumentation is undermined with more and more farmers deciding in favor of GM crops. What these activists do as a countermeasure is to take advantage of individual farmers who have suffered crop failures due to drought or other problems and depict these failures as outcomes of Bt cotton adoption. The misfortune of individuals often makes good stories—so good that one easily forgets about the millions of other farmers who are extremely satisfied with Bt cotton and how the technology has improved their lives.

Another hoax is the story of "Monsanto's terminator gene" which was created by international NGO networks and propagated in India by Vandana Shiva and Navdanya (Herring, 2010). The terminator gene is a genetic use restriction that renders second generation seeds sterile. The narrative says that this gene was incorporated into Bt cotton seeds sold in India with the intention to increase corporate control of the seed sector and enforce farmers to buy new seeds every year. In addition to farmers' dependence, Shiva also declared a huge environmental risk. In one of her publications she wrote: "the possibility that the terminator may spread to surrounding food crops or to the natural environment must be taken seriously. The gradual spread of sterility in seeding plants would result in a global catastrophe that could eventually wipe out higher life forms, including humans, from the planet" (Navdanya, 2004, p. 9). But this story had several problems. First, terminator seeds or outcross-ing transgenes cannot spread in the environment because sterile plants would not be able to reproduce. Second, Bt cotton does not contain the terminator gene and never did. Third, while a genetic use restriction technology was developed jointly by the United States Department of Agriculture (USDA) and Delta & Pineland in the 1990s, this technology, which was dubbed "terminator" by a Canadian NGO, was never com-mercialized and used anywhere in the world because of public resistance. Nevertheless, the NGO campaign targeting terminator seeds was very successful. Many people all over the world believe that GMOs come with genetic restrictions that prevent farmers from saving and replanting seeds (Herring, 2010). In reality, such genetic use restrictions do not exist in commercialized or pipeline GM technologies.

NGO Reactions to Studies Showing Positive Impacts

NGO activists and their communication teams are not only creating their own narratives about negative impacts of GM crops, sometimes invent-ing stories completely and sometimes building on questionable scientific results, but often they also react to published studies showing positive GMO impacts. Let me report about some own experience I made. This experience is consistent with observed NGO reactions to many other studies.

Whenever my research team published a paper in an academic jour-nal showing positive effects of GM crops for farmers or consumers, and the results were not picked up by the public media, anti-biotech NGOs took note but remained silent. This is a first indication that NGOs are not interested in scientific evidence as such. However, whenever a paper received media coverage, thus bearing the risk that the broader public could notice, NGOs were extremely quick in their reaction to denounce the results. This happened several times during the last couple of years, mostly related to my work on the impacts of Bt cotton in India (chapter 4). NGOs know that scientific studies showing significant benefits for smallholder farmers in terms of lower pesticide use, higher yields, and higher family incomes can undermine their narratives when the results remain undisputed. I hardly ever received any direct communication from NGO representatives with a request to respond to questions or counter-arguments, which I take as another indication that they are not really interested in scientific dialogue. Instead, their views, critical comments, and false allegations are posted on their Internet blogs, in open commentary sections of news websites, or other social media channels. These types of reactions are not only posted in one location, but the same arguments with identical wording often appear almost instantaneously on dozens of websites around the world. NGOs have very tight and well-functioning international networks.

What are the types of arguments used by NGOs in their reactions to positive impact papers? Typically, critical points are raised with respect to sample size and sample selection, statistical approach, interpretation of results, conclusions reached, and inconsistency with the real situation on the ground. The lists of points raised tend to be long, so the non-expert reader of the blog must get the impression that the study was very poorly executed. Discrediting the study authors in the wider public is clearly the intention. However, to the expert reader it becomes obvious that the commentators are unfamiliar with the research methods and have not even read the original paper in full. It suffices that the results are not in line with their narratives. To back the alleged inconsistency with the situation on the ground, NGOs often cite their own publications or quote selected voices of local farmer representatives disapproving the technology. Closer scrutiny of who these farmer representatives are often reveals that these are NGO activists themselves.

As for any research, if there is justifiable criticism after publication of a study in a peer-reviewed journal, the usual procedure in the science community is to write a formal comment, which will also be peer-reviewed and published in the same or a different journal if deemed relevant. None of the comments by NGOs to my papers on Bt cotton were published in an academic journal. The comments would likely not have passed the peer-review process, but I assume that this route was not even tried. Posting comments in various Internet forums without any quality control is not only easier but also much more effective in terms of global outreach. The wider public does not differentiate between peer-reviewed research findings and unsubstantiated claims. The Internet is also useful for the statement that this is yet another publication of the same author whose papers were so harshly criticized in the past. A simple Internet search suffices for the layperson to judge that this claim is apparently true, as the same criticism with respect to earlier papers is found on so many websites and Internet blogs.

Another standard argument by anti-GMO activists is that studies showing positive impacts were influenced by the biotech industry and funded through company money. Vandana Shiva asserts that all research from my group is "manipulated" and concludes: " . . . Qaim represents Monsanto & Co. Every 'study' done by him is public relations for Monsanto" (Shiva et al., 2011, p. 153). In reality, my research on Bt cotton and other GMOs was entirely funded by grants from public sector and philanthropic organizations and was never influenced by private sector interests. The industry money argument is popular among NGOs whenever research results do not please them. Most anti-biotech NGOs do not care whether or not this is true. What I find interesting is that NGOs never use this argument when a study was funded by the organic food industry. Not to be misunderstood, I do not deny that industry influence can happen in all areas of science. Hence, a critical eye on potential conflicts of interest is important when interpreting research results. But accusing all studies to be biased whenever the results do not fit the own ideology is perhaps too simplistic.

NGOs have also made every effort to block and discredit the research on Golden Rice. This may surprise, because the development of Golden Rice is a humanitarian project that does not pursue any industry objectives and—once released—could save thousands of lives every year (chapter 5). But this is exactly the problem. Many in the anti-biotech movement perceive Golden Rice as a Trojan horse that, if made widely available, could fundamentally alter the public debate about GMOs. Greenpeace is one of the most prominent opponents of Golden Rice. In a recent publication they assert that Golden Rice is an ineffective tool to combat vitamin A malnutrition and poses risks to human health and food security. Furthermore, they warn that the GM rice would lead to genetic contamination of rice landraces in Asia. Greenpeace concludes that "spending even more time and money on golden rice development is not only environmentally irresponsible, it is also a disservice to humanity" (Greenpeace, 2013, p. 9).

Greenpeace's fight against Golden Rice already started 20 years ago. In the mid-1990s, they intercepted GM rice seed that was sent for research purposes from the Swiss Federal Institute of Technology in Zurich to

IRRI in the Philippines. At the same time, Greenpeace also started to exert their anti-GMO influence on local NGOs in the Philippines (Aerni, 2014). In 2001, after publication of the Golden Rice proof of concept, Greenpeace pointed out that the low concentrations of beta-carotene present in the grain meant that people would need to consume 12 times the normal daily intake of rice in order to obtain the recommended dietary allowance of vitamin A.

In 2009, results of a trial carried out with adults in the United States were published showing that the beta-carotene in Golden Rice is highly bioavailable to humans (Tang et al., 2009). Greenpeace promoted criticism that the adult subjects in the US experiment were not vitamin A deficient, so that the results would be of little practical value. In 2012, results of an additional trial carried out with vitamin A deficient children in China were published in the *American Journal of Clinical Nutrition*, confirming the high bioavailability of the beta-carotene (Tang et al., 2012). The same study also showed that a portion of 100–150 g of cooked Golden Rice (50 g dry weight) can provide 60 percent of the recommended vitamin A intake for children. Following this publication, Greenpeace issued a press release condemning the use of a GMO-crop with Chinese children as "guinea pigs of American researchers" (Dubock, 2014). In fact, several of the researchers involved in the study were Chinese, and the experimental protocol was in full agreement with Chinese legislation. Nonetheless, Greenpeace's press release and the wide international media coverage stirred concern among Chinese authorities. Three of the local scientists were threatened and eventually sacked for their involvement in this research (Eisenstein, 2014). A subsequent review by Tufts University, the American organization involved in this research, confirmed the validity of the study results, but identified concerns with the informed consent process, especially inadequate explanation of the GM nature of Golden Rice. However, the institutional review board of Tufts University had previously approved the wording of the informed consent form to be used with the experimental subjects and their parents (Dubock, 2014). As a result of the debate that had been instigated by Greenpeace, the *American Journal of Clinical Nutrition* decided to retract the paper in July 2015. In the retraction notice, the journal did not question any of the results but mentioned insufficient documentation of the informed consent process.

In 2013, a field trial with Golden Rice was destroyed in the Philippines by a large crowd of people (Alberts et al., 2013). The trial was carried out jointly by the government's Philippine Rice Research Institute and IRRI. NGOs portrayed the vandalism as an act of angry farmers concerned about the risks of GM rice. However, it turned out that the incident was planned and implemented by NGO activists themselves. Greenpeace denied any involvement. But a local NGO with the name MASIPAG was identified by the local Department of Agriculture to have been involved. Greenpeace has cooperated with MASIPAG in the Philippines since the 1990s (Aerni, 2014).

Wider Implications of NGO Narratives

The NGO narratives about GMOs have dominated the agricultural biotech debate since the 1990s and have been the most decisive factor in forming public opinions about this technology around the world. The globally most influential NGOs in the GMO debate are Greenpeace and Friends of the Earth, which have their headquarters in the Netherlands, local offices in many other countries, and a huge network of partner NGOs on all continents. The European-based international NGOs often claim that they represent the voices of the poor in developing countries and help local NGOs to make these voices heard internationally. However, in the GMO debate the direction of influence is mostly the other way around. That is, Western NGOs set the agenda and tell their local partner organizations in developing countries what to protest against.

At the World Summit on Sustainable Development in Johannesburg in 2002, Friends of the Earth and other European NGOs coached their African partners to sign an open letter warning that GM foods could cause allergies, chronic toxic effects, and cancers (Paarlberg, 2014). A stakeholder survey carried out by Philipp Aerni in different developing countries showed that most of the local organizations opposed to GM crops were funded by European NGOs and development agencies. A MASIPAG representative in the Philippines mentioned that focusing his advocacy work on resistance to GMOs ensures that he is getting invited to Europe to talk about the farmer victims of multinational biotech companies. A local Greenpeace representative in Mexico mentioned that she once tried to convince the headquarters that Mexico has more important local issues that affect people and the environment than GMOs. "She was rebuffed and told to look for another job if she could not put up with the Greenpeace agenda" (Aerni, 2014, p. 263).

A powerful NGO network that claims to represent around two hundred million peasant farmers around the world is La Via Campesina. La Via Campesina may sound like a Latin American social movement, but is an organization that was founded in Belgium in 1993. The network grew over time with significant financial support from Europe. Today, La Via Campesina comprises 164 local organizations in 73 countries in

Europe, Africa, Asia, and the Americas (La Via Campesina, 2015). Via Campesina launched the idea of "food sovereignty," which it defines as the right of peoples to healthy and culturally appropriate food produced through sustainable methods and their right to define their own food and agriculture systems. What exactly that means in practice is subject to wide interpretation, but what it does not mean for Via Campesina becomes clear from its protest actions and events. Via Campesina says no to trade liberalization, multinational corporations, the World Bank and other international development banks, GMOs, hybrid seeds, agro-chemicals, and market-based farming. From my own research experience with peasant farmers in developing countries I have my doubts that all of them feel well represented by an NGO network that tries to perpetuate low-tech subsistence farming.

The media likes to pick up NGO narratives about GMOs, but politicians also tend to listen very carefully. Even politicians who do not agree with NGOs consider their lobbying efforts as an important source of information because it is assumed that these civil society organizations represent public opinions. In reality, NGOs shape public opinions much more than representing them. The unofficial EU moratorium on GM crop approvals between 1999 and 2003 (chapter 6) was clearly the result of NGO influence. In Zambia in southern Africa, Greenpeace told government officials that their European export market for organic food would collapse if GMOs were let into the country. The US organization Genetic Food Alert warned Zambia in 2002 of the unknown health risks of consuming GM foods, the British organization Farming and Livestock Concern alleged that GM maize could form a retrovirus similar to HIV (Paarlberg, 2014). These and other claims led the government of Zambia in 2002 to reject 35,000 tons of food aid donated by the United States through the World Food Program, because the shipment contained GM maize. In 2002, more than 2 million people in Zambia suffered from acute food shortages.

In 2004, the Food and Agriculture Organization of the United Nations (FAO) brought out its annual flagship publication, *The Sate of Food and Agriculture*, under the theme "Agricultural Biotechnology: Meeting the Needs of the Poor?". This publication reflected the state of the art about GMO impacts at that time and discussed potentials and issues for developing countries. Scientists commended this report as a well-researched and balanced document. But most NGOs were not amused that a UN organization saw some potential in GM crops to benefit poor farmers. Several hundred NGOs wrote an open letter to the FAO Director General in which they accused the report to be a tool of the biotech industry. The FAO response to the open letter was quite accommodating (Evenson and

Raney, 2007). Since then, FAO has hardly touched the topic of GMOs anymore in its official publications.

Also in 2004, Greenpeace activists destroyed a field trial with GM papaya in Thailand, causing a countrywide moratorium on all field testing of GM crops (chapter 4). In 2007, the opposition by Greenpeace and Friends of the Earth contributed to a temporary ban on the cultivation of GM crops in most states of Australia. In 2011, Greenpeace was involved in destroying a field trial of Bt eggplant in the Philippines. In 2013, the Philippine Court of Appeals followed a petition by Greenpeace and MASIPAG to ban any further field trials with Bt eggplant in the country (Laursen, 2013). The Court's decision was justified using the retracted Séralini study as evidence. In 2012, the same Séralini study was also the main reason for Kenya's government to ban the imports of any GM food into the country.

Even in India, where farmers made very positive experience with Bt cotton, anti-biotech pressure groups continue with their political lobbying. The Bt eggplant ban in India in 2009 (chapter 4) was clearly influenced by Western NGOs. After many years of testing, Bt eggplant had been declared safe for the environment and human health by the Indian biosafety and food safety authorities. But the Minister of the Environment banned Bt eggplant, claiming health risks based on a 2008 paper by Gilles-Eric Séralini. In that paper, Séralini concluded that the consumption of Bt eggplant could result in human organ failure and potential death. This paper was not peer-reviewed, nor was it based on any experiments carried out with Bt eggplant. Instead, earlier work of Séralini with Bt maize was cited as evidence. Séralini's paper on Bt eggplant cites Greenpeace India as a source of funding; his earlier work on Bt maize was sponsored by Greenpeace Germany. The EFSA GMO panel had reviewed this work and concluded that the claims regarding negative health effects are not supported by the data (Herring, 2015).

The Bt eggplant ban in India is still in place and is only one example for the controversies that anti-biotech campaigners have stirred in a country where millions of smallholder farmers have improved their livelihoods through Bt cotton adoption. But these benefits are hardly recognized by the urban non-farm population that is flooded with negative propaganda about GMOs. In 2012, a high-level panel of the Indian Parliament launched a report concluding that GM crops would harm rather than benefit Indian agriculture. This report, which built heavily on Vandana Shiva's claims about suicides and other alleged negative social consequences of Bt cotton for smallholder farmers, called for an immediate halt of all GMO field trials in the country (Bagla, 2012). A technical expert committee was appointed to further review the evidence with inputs from scientists before further political decisions are made. While the ban on GMO field trials was not implemented, the situation for new GM crop approvals in India remains difficult due to continued NGO lobbying efforts (Krishnan, 2015).

Also in China, a country that has heavily invested in public sector research on agricultural biotech since the 1980s, NGO campaigns have a strong influence on GMO policies. Greenpeace claims to have been largely responsible for the fact that GM rice has not yet been commercially approved in China (Greenpeace, 2015).

Beyond their strong influence on public attitudes and policy decisions in individual countries, NGOs have a systematic influence on international biosafety agreements and policymaking processes (Falkner, 2007). Since the 1990s, the UN approach of discussing international issues and reaching agreements has explicitly become more open for groups representing the civil society. NGOs are considered to represent public views. While their role within the UN system remains consultative, NGOs have gained much greater access to international decision-making arenas that were previously the sole domain of the member states (Panjabi, 1997). The negotiations on the Cartagena Protocol for Biosafety were one of the early UN forums where NGOs had a significant say. They lobbied vehemently in favor of the precautionary principle and its odd interpretation, where trade in GMOs is likened to the transboundary movement of hazardous wastes (Paarlberg, 2008). As discussed in chapter 6, the Cartagena Protocol is the international agreement that provides the framework for biosafety policies in all countries that have ratified the document.

NGOs also play a growing role in influencing scientific policy advice. Several national ministries and international organizations include NGO representatives as members of their scientific advisory panels. In 2005, the World Bank and the UN initiated a process to analyze the state of agricultural research and how new technologies can contribute to sustainable development. This process became known as the International Assessment of Agricultural Science and Technology for Development (IAASTD) and was eventually driven by NGOs to a significant extent (Stokstad, 2008). No surprise that in the final IAASTD report hardly any potential was seen for GM crops (IAASTD, 2009). Since its publication, the IAASTD report is touted by NGOs and many in the wider public as a consensus paper among international scientists. But it is not. The report is rather a consensus paper of international NGOs.

In 2014, Greenpeace and other NGOs had conducted a concerted campaign against Anne Glover, the chief science advisor of the EU Commission's President at that time. Anne Glover, a professor of cell biology, had a very science-based approach and openly supported GMOs, which is why she was disliked in the NGO community. In November 2014, the incoming President of the EU Commission, Jean-Claude Juncker, refused to renew Glover's contract and abandoned the position of chief science advisor altogether.

NGO narratives and lobbying efforts have directly caused delayed and denied GMO approvals, technology bans, and a shrinking role of science in national and international policymaking. NGOs have also affected R&D priorities and trends in the plant sciences more broadly, possibly with profound long-term implications. Several biotech companies have moved their research out of Europe. More than a few public research groups have stopped the development of GM crops. Especially young plant biotech researchers are often intimidated by the emotional opposition they experience when mentioning their field of research in the public. Many prefer to keep silent about their work in private life, others have decided not to enter this field of research in the first place.

The power of NGOs in shaping the GMO debate, societal climates, and technology policy decisions is frightening given that these organizations are not democratically elected and not really accountable to anyone.

Is There Scientific Consensus?

In the previous sections, I have discussed how much NGO positions on GMOs differ from the empirical evidence and scientific opinions. Does this imply that there is scientific consensus about GMO potentials and risks? No. As for almost every scientific topic, there are heterogeneous views. Not all scientists agree that climate change is caused by human activity. There is not even a consensus on whether there is global warming at all. There is a lively debate about the health risks of mobile phones and wireless connections. Researchers also dispute whether or not the use of biofuels contributes to global hunger, or whether the promotion of kitchen gardens is a useful approach to improve dietary diversity in developing countries. Controversy is useful and important to advance science and promote sustainable development, but denying facts because they do not match own ideologies or inventing extremely unlikely risks is not.

It is useful to differentiate between different aspects of GMOs and take stock for which of these aspects there are established facts, for which there is almost universal consensus, and for which scientific views are more diverse. Established facts exist for some of the benefits of GM crops. Numerous scientific studies have shown that the adoption of commercialized GM crops has led to higher yields and higher profits for farmers in developed and developing countries. Bt crops have contributed

to significant reductions in the use of chemical insecticides with con-
comitant environmental benefits. There are individual studies that do
not find benefits in specific situations. This is unsurprising, as not every
technology is useful in every situation. For instance, in low pest pressure environments insect-resistant crops are not required. Meta-analyses
that combine the results from all individual studies clearly establish that
GMOs cause sizeable benefits on average. This evidence was reviewed
in chapter 4. Beyond the already commercialized GM crops it is also
an established fact that genetic engineering helps to develop other crop
traits with large potential to contribute to sustainable development and
climate change adaptation, such as disease resistance, drought tolerance,
salt tolerance, and enhanced nutrient use efficiency. All of these new GM
traits have already been tested successfully in the field, as was discussed
in chapter 5.

Very wide scientific consensus also exists for the point that GM crops
are not *per se* more risky than conventionally bred crops. This has been
established in thousands of risk-assessment studies over the last 30 years.
In fact, due to the strict regulatory requirements for GM crops, these
crops have been evaluated for risks much more comprehensively than
conventionally bred crops and most other agricultural technologies.
Based on this compelling evidence, national science academies from
many countries have concluded that genetic engineering is not associated
with new kinds of risks. This does not mean that there are zero risks, but
the risks are related to the crop traits, not the breeding process. One can
always question individual studies and researchers. But science academies
are organizations that comprise a large number of independent top scientists from various disciplines. Academy members are elected based on
scientific merit. Hence, these academy statements must be considered as
reviews of the best scientific knowledge available.

It should be mentioned that there are a few institutes and networks
of scientists that challenge the evidence available and postulate that GM
crops are dangerous. One such network is the European Network of
Scientists for Social and Environmental Responsibility (ENSSER).
ENSSER claims to be a network of independent scientists working for
the public good. However, its mission statement reveals that ENSSER
has a clear agenda, namely "the protection of the environment, biological diversity and human health against negative impacts of new technologies and their products" (ENSSER, 2015). Its main activities are
related to hyping controversial studies, such as the Séralini papers, orga-
nizing media events, and directly lobbying politicians. The Committee
for Independent Research and Information on Genetic Engineering
(CRIIGEN), which was co-founded by Gilles-Eric Séralini in 1999, has

a very similar agenda. There are also several NGO campaigners who founded their own institutes with scientifically sounding names. Already in the 1980s, Vandana Shiva established the Research Foundation for Science, Technology, and Ecology in India.

ENSSER, CRIIGEN, and like-minded organizations fit the definition of what Marcel Kuntz has called "parallel science" of NGO advocacy groups. These groups like science when it confirms their views. When it contradicts them, rather than changing their minds, they often prefer to change the science to fit their ideology (Kuntz, 2014). In the broader science community, these institutes and networks are seen as low-profile organizations with a heavily biased agenda, but because of the very close association with NGOs their public outreach and policy influence is significant. It is certainly true that individual studies arguing against the mainstream are important to advance the knowledge and identify unexpected phenomena. But scientific standards still need to be maintained. As independent reviews have demonstrated, the studies promoted by these parallel science organizations regularly suffer from serious methodological flaws. The problem is that the flawed papers are widely hyped in the Internet, while the reviews by independent science panels are hardly recognized in the wider public.

Not counting such parallel science, there is broad consensus on the large potentials and manageable risks of GM crops. This means there is almost universal agreement that a ban on GMOs is not scientifically justified. Beyond this bottom line, the views of scientists are more diverse. There are some who do not like specific traits, such as herbicide tolerance, which—if not properly used—can contribute to lower system diversity. Others are concerned about patents and monopolies. Some also think that other technologies are more suitable for smallholders. But such multiplicity of arguments on specific aspects is normal and also observed for other agricultural technologies and approaches. For instance, not everybody agrees that organic farming is a useful and superior approach, without necessarily calling for a ban on organically produced foods.

Why Is the NGO Anti-Biotech Propaganda So Successful?

In the previous sections I have argued that there is a large protest industry against GM crops, building on narratives that have become public truths even though they are inconsistent with the scientific evidence. These narratives were created and are constantly nurtured by anti-biotech NGOs with support from the organic farming and food industries and a few other groupings that benefit from the GMO opposition.

But an important question is why NGOs are so successful in influencing public attitudes.

There are very different types of NGOs with different activities, ranging from the implementation of local charity and environmental improvement projects to broader educational and awareness creation campaigns. Some of the large international NGOs—such as Greenpeace International and Friends of the Earth International—concentrate primarily on campaigning and political lobbying. They have very large budgets available for their public relations (PR) work, much larger than any public sector research organization. They also have experienced PR professionals employed who know how and when to successfully launch international media campaigns. International NGOs also have a huge network of local offices and partner organizations in numerous countries, thus ensuring that messages are widely spread and translated into multiple languages. And the messages of NGOs are easy to understand and often intuitively plausible, because they confirm widely held prejudices. When truthfulness is not a criterion, it is simple to create Twitter-like slogans that are hugely effective globally.

It could be argued that private multinational companies also have huge PR budgets and global networks to build on, so that there is basically a level playing field. However, there is one huge difference of crucial importance for the success of PR strategies, namely trust. Most people do not trust large private companies, but they trust NGOs. NGOs are considered as inherently good, with idealistic motives and no profit incentives. They are perceived to fight for the environment, for the economically disadvantaged, and against corporate greed and corrupt governments.

Social psychologists differentiate between two types of trust, confidence and social trust (Siegrist et al., 2007). Confidence is based on own understanding and experience, whereas social trust is the willingness to rely on the judgment of others whom we trust. Own understanding of GM technology is hardly possible for the wider public, because the underlying science is complex. Own experience is also lacking, especially in Europe where no GM crops are approved for cultivation in most countries. But also elsewhere, most of the GM crops commercialized so far are used for feed and fiber production, so that food consumers have little immediate familiarity with GMOs. In this situation, most people have not developed own confidence but resort to social trust. People trust others whose values are compatible with their own goals. Social experiments have shown that people who have to make judgments with limited information take cognitive shortcuts (co-called heuristics), for instance by relying on the judgment of others with whom they share common values (Kahneman, 2011). NGOs with their perceived idealistic motives therefore enjoy a lot of trust; their judgment on GMOs is taken at face value. Heuristics are not necessarily rational, but they deliver solutions for the individual with minimum cognitive effort (Busch, 2010). In this context, it does not matter that environmental NGOs may not be the groups with the best scientific expertise on GMOs.

NGOs know about these psychological effects and try hard to avoid differentiated views on GMOs. They divide the world into good and evil, this bipolar approach serves their purpose best. GMOs are stigmatized as evil, but in order to block the possibility that other groups may also enjoy social trust, it is important for NGOs to also depict all those groups and individuals as evil who see some potential in GM crops. Biotech companies are certainly evil, as they are seeking profit at the expense of consumer health, the environment, and resource-poor farmers. Hence, it is opportune to simply portray public sector researchers who show positive GM crop impacts as being bought by industry. Journalists who report positively about GMOs are also accused of being bribable. Even UN organizations like the FAO have been villainized by NGOs to represent industry interests, as explained previously. In this bipolar world view advocates of GMOs can no longer be perceived in terms of value similarity.

Another important psychological aspect in the public GMO debate is fear. Fear is a crucial element in human behavior, which has evolutionary roots. Historically, life was always risky for humans. So the ability to react promptly to fear was essential for survival. This makes fear much more powerful than reason to explain behavior (Rosen and Schulkin, 1998). NGOs exploit this effect by creating fears and thus arousing rejection and defensive action. Can there be anything more fearful than GMOs that cause cancer and allergies, contribute to antibiotic resistance, drive farmers into suicide, and cause environmental disaster?

Too Little Counterweight in the Public Debate

NGOs are the dominant force in shaping public opinions about GM crops, but they are only one group of actors. What is the role of other relevant actors and why is there not a more effective counterweight to the NGO narratives? The answer is mixed. Some groups benefit from biased public attitudes about GM crops, others try to contradict but are not very effective, and yet others simply keep quiet because of intimidation and fear to jeopardize their public reputation. The organic food industry is one grouping that clearly benefits from public resistance to GMOs. Organic farmers' associations and food chain representatives join forces with NGOs to depict GMOs as dangerous and undesirable. Organic farming is seen by many in the wider public as the role model of sustainable agriculture and healthy food production. Due to its perceived moral integrity, the organic food industry enjoys a lot of social trust.

The group of researchers is much more diverse. Some researchers are not particularly active in public communication, as this is time-consuming and not rewarded in scientific incentive systems. Publication in academic journals is scientifically much more rewarding, but these journal publications are neither accessible nor comprehensible to the wider public. A few researchers do communicate publicly, but whenever they have positive things about GMOs to report, NGOs try hard to publicly discredit them. Moreover, unlike the NGO propaganda, scientific statements are often not suitable for Twitter-like slogans. Researchers have an academic reputation to lose, and without a few ifs and buts short statements could be misinterpreted. However, with many ifs and buts messages are not particularly effective in building social trust. And then there are the contradicting statements by "parallel science" organizations. Statements by science academies are useful to clarify state-of-the-art knowledge within the scientific community, but the wider public cannot easily differentiate between more and less respected science organizations. Ultimately, in the eyes of the public, scientific results remain heterogeneous and confusing.

Journalists also play an important role in science communication. As explained earlier, fear stories about new technologies are more lucrative for the media than stories about benefits. Journalists of more serious media houses try to report in a balanced way, but this can also be problematic. It has become quite common that media reports about new scientific studies end with an NGO statement about the plausibility and implications of the results. While balanced perspectives are laudable in general, scientific results must not become a matter of democratic lobbying and voting. However, journalists who report about GMOs without a critical undertone have to expect NGO bashing in the social media and through letters to the editor. Only very experienced correspondents with a lot of expertise in the subject area will dare to do so. Most journalists are not so deep into one particular subject, which is why they prefer to rather stay politically correct.

And what about politicians? They also contribute to the success and persistence of NGO narratives, or they keep silent. Green parties in particular like to propagate NGO stories about GMO risks and lacking benefits. In some countries, green parties have even made the goal of banning GMOs one of their main topics in electoral campaigns. Speaking out against GMOs is very popular in several European countries and hence a welcome possibility to collect additional votes. Even conservative parties partly ride on this train, often with the argument that they cannot ignore that the majority of the population is against GMOs. Many politicians do not share these negative attitudes toward GMOs, as I know from several private discussions. But very few would challenge the anti-biotech propaganda in public, as this would surely lead to NGO accusations of being bribed by industry. So these politicians rather remain quiet on the topic. At least in Europe, GMOs are not of sufficient immediate importance to jeopardize losing a potentially sizeable number of votes. I acknowledge that there are a few politicians who speak out in favor of GMOs in public, but these are notable exceptions.

Are NGOs Pursuing Only Idealistic Motives?

NGOs claim to fight for the environment, for the rights of the poor and disadvantaged, and against corporate and political greed. These are noble motives. In the wider public, NGOs are perceived as honest warriors for the good, completely independent of political and profit objectives, which gives them an almost sacrosanct status in the eyes of many. But are NGOs really so idealistic? Do they believe their own narratives about GMOs, or do they use them as instruments for a deeper cause? What are the real motives of their biotech opposition? This last question is not so easy to answer. The arguments brought forward by NGOs are manifold and fundamental. But NGO representatives are often not interested in a detailed discussion. Each time evidence-based logic prevails over a particular anti-GMO position, another often unrelated objection is raised (Dubock, 2014).

NGO opposition against modern agricultural biotechnology started in the mid-1990s. The early pressure groups against GMOs were those that had their roots in the environmental movement of the 1970s. At that time, these pressure groups criticized the negative environmental effects of the Green Revolution. By assuming that the use of GMOs would lead to a repetition of the mistakes of the Green Revolution, they were able to shape the environmental risk narrative to a great extent (Aerni, 2014). Activists in the 1990s were particularly concerned about the potentially irreversible effects of GM crops on biological diversity. At that time, GM technology and the potential behavior of transgenes were less well understood. Some scientists shared the environmental concerns, so a somewhat cautious approach to releasing GMOs into the environment was justified. Already at that time, NGOs exaggerated the environmental risks beyond what was scientifically plausible, but this is normal for campaigning organizations. This was also the time when many governments were considering how to regulate GMOs. Especially in Europe, NGOs successfully influenced governments to regulate GM crops separately from other breeding technologies, and with much higher scrutiny. By nurturing fears about GMOs, NGOs gained a new topic with large potential to generate financial revenues through private donations from concerned individuals, government grants, and contributions from various foundations.

Campaigning against GMOs became a lucrative business for European and American NGOs and allowed them to grow in size and international outreach. Especially in developing countries, the number of local NGOs grew significantly in the 1990s, many with an interest in agricultural topics. Western governments were happy to support the development of such local organizations with the objective to strengthen civil society in new and sometimes fragile democracies. Local NGOs in developing countries also became interesting partners for European and American NGOs in their efforts to spread fears about GMOs globally. Fear-mongering did not stop when an increasing number of scientific studies showed that the initial concerns about the environmental and health risks of GMOs had been grossly overblown. Rather, fear-mongering was broadened to social risks of GMOs, including new narratives about the exploitation of smallholder farmers and corporate dominance of the food chain. NGOs framed the opposition against GMOs as a fight of David against Goliath: the honest warriors for the good against the mighty and greedy agri-biotech multinationals. For several NGOs, the fight against GMOs and Monsanto has become an important element of their global brand value.

In the 2000s, scientific evidence accumulated showing that GM crops do not lead to social catastrophe but instead significantly benefit smallholder farmers in developing countries. In addition, more and more research revealed that GM crops can also contribute to more environmentally-friendly agricultural production, especially through reductions in chemical insecticide use. However, by that time the bipolar world view was sufficiently established, and the NGO narratives had become public truths. There was no need to change the storylines and jeopardize a very lucrative business model. At least on issues of agricultural biotechnology, it is increasingly clear that NGOs are not really interested in environmental and social improvements. They are much keener on keeping public fears alive, as these fears ensure a steady and growing flow of revenue. From that perspective, anti-biotech activist organizations may not even be interested in a complete global ban on GMOs, as entirely "solving" the issue would likely also have an unfavorable effect on donations (Graff et al., 2009).

To understand the financial magnitude of the NGO sector and its campaigns against GMOs a few numbers may be useful. Apel (2010) estimated that in the mid-2000s Friends of the Earth and its member organizations and affiliates had received over 600 million US$ annually from the EU public sector, including the European Commission and EU member country governments. Greenpeace has a different approach; it does not accept government grants and relies on donations alone. Nonetheless, its annual budget is very sizeable. In 2011, Greenpeace International and its affiliated national and regional organizations had global revenues of 336 million US$ (Nisbet, 2013). Not all of this money is directed to anti-GMO campaigns, but a significant proportion is.

Changing positions on GMOs and admitting that earlier statements were wrong does not seem to be an option for large NGOs. Such admission would not only reduce revenues from donations targeted at biotech but could also seriously damage reputation and social trust, which are very important for the success of all sorts of campaigns. NGOs also exercise considerable peer pressure on other organizations in the protest industry that might break out and disturb the harmony. Individual NGOs—such as Oxfam and the World Wide Fund for Nature (WWF)—seem to have softened their positions on GMOs very slightly, but more than this does not seem to be possible without being stigmatized by the crowd of other NGOs.

Personal withdrawal from the organizations seems to be the only option for individuals who disagree with the extreme positions of environmental NGOs and their unwillingness to compromise. There are several cases of former NGO campaigners who publicly announced their withdrawal and now speak in favor of GMOs. Patrick Moore was one of the founding members of Greenpeace in the early 1970s. He was president of Greenpeace Canada, and from 1977 also president of the Greenpeace Foundation, what later became Greenpeace International. In 1986, he abandoned the organization and later criticized Greenpeace's positions as unscientific and ideologically biased. Moore's dropout was before GMOs became a major topic for Greenpeace, but he recently also condemned the organization's approach to agricultural biotechnology (Moore, 2010). In 2013, Patrick Moore started the "Allow Golden Rice Now" campaign. Unsurprisingly, Moore is now accused by Greenpeace as a paid representative of the biotech industry.

Another interesting case is Mark Lynas, a British environmentalist who for many years campaigned against GMOs with Greenpeace and the UK Soil Association, an organic trade group. In January 2013, Lynas confessed his conversion from an opponent to a GMO supporter by addressing the Oxford Farming Conference with the following words:

> I want to start with some apologies. For the record, here and upfront, I apologize for having spent several years ripping up GM crops. I am also sorry that I helped to start the anti-GM movement back in the mid-1990s, and that I thereby assisted in demonizing an important technological option which can be used to benefit the environment. As an environmentalist, and someone who believes that everyone in this world has a right to a healthy and nutritious diet of their choosing, I could not have chosen a more counter-productive path. I now regret it completely. So I guess you'll be wondering—what happened between 1995 and now that made me not only change my mind but come here and admit it? Well, the answer is fairly simple: I discovered science, and in the process I hope I became a better environmentalist. When I first heard about Monsanto's GM soya I knew exactly what I thought. Here was a big American corporation with a nasty track record, putting something new and experimental into our food without telling us. Mixing genes between species seemed to be about as unnatural as you can get—here was humankind acquiring too much technological power; something was bound to go horribly wrong. These genes would spread like some kind of living pollution. It was the stuff of nightmares. These fears spread like wildfire, and within a few years GM was essentially banned in Europe, and our worries were exported by NGOs like Greenpeace and Friends of the Earth to Africa, India, and the rest of Asia, where GM is still banned today. This was the most successful campaign I have ever been involved with. (Lynas, 2013)

CHAPTER 8

CONCLUSIONS

In spite of notable progress in global hunger and poverty reduction over the last few decades, way too many people in developing countries are still not able to satisfy their basic needs. Close to eight hundred million people are undernourished and do not have sufficient access to calories, most of them living in Asia and Africa. Urbanization tendencies notwithstanding, around 75 percent of the undernourished people reside in rural areas where they directly depend on agriculture as a source of income and employment. In addition to insufficient calorie intakes, micronutrient malnutrition is a serious issue. Around two billion people suffer from deficiencies in specific minerals and vitamins. These forms of malnutrition are a humanitarian disaster. They contribute to numerous infectious diseases, involve physical and mental retardation, and are the leading causes of child mortality in developing countries. Undernutrition and micronutrient malnutrition also cause huge economic costs, obstructing growth and development. Addressing these problems needs to be on top of the global development agenda.

Apart from acute crisis situations, lack of food is currently not the main reason for the observed widespread chronic hunger. Significant investment in agricultural R&D during the twentieth century enabled technological progress such that growth in global food supply outstripped growth in global demand. However, public sector investments in agricultural R&D were downsized in the 1990s. This was largely due to the sizeable productivity increases between the 1960s and 1980s and the false conclusion that further hunger reduction would not depend so much on agricultural growth anymore. Due to time lags between research and measurable effects on agricultural production, the repercussions of the declining R&D investments have only been felt since the 2000s. Yield growth in major food crops today is much lower than it was in previous decades, while demand growth has continued to be strong. As a result, global food prices have shown an increasing trend during the last 10 to 15 years. In addition, the tighter supply situation has contributed to several recent food price crises. These developments make it even more difficult for poor people to access food of sufficient quantity and quality.

Growth in food demand will continue to be strong over the next few decades, due to rising population numbers and rising incomes in most developing countries. Feed demand will also increase significantly because the rising middle classes in developing and emerging countries tend to consume more meat and dairy products. Finally, agriculture is increasingly seen as a source of bioenergy and other renewable resources for use in industry and construction. But land, water, and other natural resources required for agricultural production are becoming increasingly scarce. And climate change will likely worsen farming conditions in some parts of the world. In addition, there is broad evidence that the agricultural production models of the last few decades have caused environmental problems and social vulnerabilities. The problems differ by region. Where land scarcity is severe, overuse of chemical inputs is often observed. Elsewhere, limited access to inputs and technologies has led to soil nutrient depletion and cropland expansion into ecologically fragile areas.

Large agricultural production increases are a sine qua non for food security and sustainable development, but the future production increases have to be more resource-conserving and need to avoid the negative environmental externalities of the past. I am very optimistic that the 9 billion or more people that will likely live on our planet by 2050 can be fed and provided with other essential goods without causing environmental disaster. Beyond 2050, projections suggest that the global population may not grow much further. Hence, it is the next few decades that are particularly challenging. Increasing agricultural production substantially and sustainably requires use and further development of the best science and technology. Too little investments in agricultural R&D, romantic views of rich country citizens about how farming should look like, and continued opposition to new technologies that were shown to be safe are all factors that entail unnecessary human suffering and environmental degradation.

GM Crops and Sustainable Development

In the introduction, I have described three goals that I see as overarching for agricultural development. The first goal is to produce sufficient food and other agricultural commodities to satisfy the needs and preferences of the growing population. The second goal is to improve the livelihoods of the people directly involved in the agricultural sector, including farmers and farm workers. And the third goal is to operate sustainably by preserving natural resources and the environment, so that the first two goals can also be achieved in the long run. There is strong evidence that GM crop technologies can and already do contribute to all three goals.

The commercial use of GM crops started in the mid-1990s. Adoption rates are uneven because not all countries approved GMOs for cultivation. Wherever these crops were approved, farmers adopted them rapidly. In 2014, around 13 percent of the global arable land was cultivated with GM crops, especially in North and South America and several countries in Asia. So far, only herbicide-tolerant and insect-resistant crops are used extensively. In chapter 4, I reviewed the available literature about impacts and showed that these crops have significantly increased yields and incomes of adopting farmers. On average, farmers in developing countries even gain more than farmers in developed countries. And consumers benefit from lower prices for agricultural commodities induced by the productivity gains. We have also seen important environmental advantages: insect-resistant GM crops have contributed to sizeable reductions in the use of chemical pesticides; herbicide-tolerant crops have facilitated the spread of no-till practices and conservation agriculture. Furthermore, the productivity increases on the cropped land have reduced area expansion into ecologically fragile areas. An estimated 25 million hectares of additional land would have been necessary to produce the same output without the productivity-increasing effects of GM technologies. Hence, the adoption of GM crops has helped to preserve biodiversity and reduce greenhouse gas emissions.

Insect-resistant cotton is so far the only GM crop widely used in the small farm sector of developing countries, especially in China, India, and Pakistan, and also in Burkina Faso, South Africa, and a few other countries. Socioeconomic research shows that GM cotton adoption has improved the livelihoods of smallholder cotton producers. There is also clear evidence that the income gains in farm households have contributed to poverty reduction, food security, and improved nutrition. Beyond the farming households, GM cotton adoption has generated additional employment, especially for female agricultural laborers, and growth in other rural sectors. In India, each dollar of direct benefits for GM cotton-adopting farmers is associated with over 80 cents of additional indirect benefits in the local economy through production and consumption multipliers (Qaim et al., 2009).

GM technologies in major food crops such as rice and wheat haven't not been commercialized up till now, largely due to anticipated problems with consumer acceptance. But, as discussed in chapter 5, GM rice, wheat, potato, cassava, sorghum, banana, beans, tomato, and several other crops have already been tested in the field with promising results. Furthermore, a number of new GM traits have been developed and tested successfully, including virus, fungal, and bacterial resistance, drought tolerance, salt tolerance, nitrogen use efficiency, and several others. Likewise, biofortified GM crops with higher amounts of micronutrients essential for human nutrition were developed and tested. Many of these new GM technologies are already at an advanced development stage and could be commercialized within the next five years. Crops that are more tolerant to biotic and abiotic stresses could produce higher and more stable yields with lower amounts of fertilizers and pesticides, thus loosening the close correlation between yield and external inputs that was characteristic for agricultural development in the twentieth century. Genetic knowledge can increasingly substitute for agrochemical inputs. Genetically hardier crops can also be an important tool to adapt to the new risks associated with climate change. In spite of these proven benefits and potentials of GM crops, public attitudes are rather negative.

GM Crops Are Not a Magic Bullet

There is large and further growing evidence that GM crops are beneficial for farmers, consumers, and the environment. However, in spite of the rapid and widespread adoption in different parts of the world, the commercial experience so far relates to a very limited number of concrete GM technologies in relatively advanced countries. Hence broad statements about the effects of GM crops in general should be avoided. It is unclear whether the results from field trials with new GM traits can be replicated when farmers start using these new technologies in their real-world production. It is also unclear whether impacts of a GM cash crop in China, India, or Argentina can be extrapolated to a GM food crop in Ethiopia or Malawi. Hence, there is still uncertainty about the implications of GM crop technologies in different situations.

The widespread adoption of new seed technologies requires that farmers have access to technical information, rural finance, seed markets, complementary inputs (if needed), and output markets. These conditions are often not fulfilled for smallholder farmers in developing countries. Especially in sub-Saharan Africa, rural infrastructure is often underdeveloped and market failures are widespread. These are also some of the reasons why the Green Revolution did not take off in Africa to the same extent as it did in Asia and Latin America. GM crops may disseminate somewhat faster than the high-yielding varieties of the Green Revolution because they are relatively easy to use and may be better adjusted to the low-input productions systems of African smallholders. Nonetheless, basic breeding capacities for introgression of GM traits into local varieties and basic seed market infrastructure are required. Without this, GM crop varieties will remain inaccessible for the poor. Of course, the same holds true for conventional seed technologies. It needs to be clear that agricultural technologies—important as they are—must not be seen as a substitute for infrastructure improvements and institutional upgrading in rural areas of developing countries. Projects that focus on the development of new seed technologies for use in regions and/or crops with underdeveloped seed market infrastructures should think about technology deployment strategies early on.

It is also important to recognize that GM crops or other seed technologies cannot substitute for good agronomy or other types of agricultural technologies. Sustainable use of improved seed varieties requires that these are integrated into suitable production systems well adapted to local conditions. Unfortunately, this is not always the case. Especially with herbicide-tolerant crops that facilitate weed control, farmers have partly switched to monocultures, growing the same crop species and using the same broad-spectrum herbicide year after year. This cannot only lead to soil degradation and rising pest infestation levels but can also entail herbicide resistance development in weed species, as observed in the Americas. These are problems that are not specific to GM crops, but they need to be addressed nevertheless. In addition to awareness building and training approaches, locally adapted rules of good agricultural practice should be established and followed. Rules concerning crop rotations can nowadays be monitored efficiently with remote sensing technologies.

Closely related to the argument on good agricultural practice and sustainable agricultural systems is the need to foster broad R&D strategies. Prioritizing recombinant DNA technologies over other breeding approaches would be wrong. For each objective the most efficient approach should be chosen. As experience shows, the development of successful new, locally adapted varieties often involves a combination of genetic engineering and conventional breeding. Similarly, prioritizing the development of improved seeds over research on improved agronomy should be avoided. There is a widespread notion that during the last 20 years, research on GM crop development received more money than other areas of agricultural R&D. At least in the public sector, this notion is clearly wrong. Compared to other research areas, GM crop development actually received relatively modest amounts of money in public sector institutions (House of Commons, 2015). No single technology can solve complex agricultural problems in isolation, which does not mean that single technologies are not essential ingredients of improved sustainable systems. Similarly, agricultural technologies cannot solve broader development problems alone, but need to be integrated into broader development strategies, together with other economic, social, and environmental interventions and policies.

GM Crop Risks and Overregulation

The public and policy debate about GM crops has focused much more on risks than on benefits of this technology. Widespread concerns about health and environmental risks have led to the establishment of complex new regulatory procedures and institutional bodies. The UN Cartagena Protocol on Biosafety has singled out GMOs as a potential threat to biodiversity and has stipulated a precautionary approach and regulatory procedures that are similar to those otherwise used for hazardous wastes. A few of the early GM crop-adopting countries—such as the United States, Canada, and Argentina—have not ratified the Cartagena Protocol, but many other countries have. This means that GM crops are more heavily regulated and tested than any other agricultural technology. In most countries, even greenhouse, confined, and open field trials need to be approved by the regulatory authorities, after submission of a comprehensive regulatory dossier. Of course, the commercial release also requires separate approval.

These regulatory procedures are not only extremely costly, they are also highly politicized, because at every step decision-makers can almost arbitrarily delay approval and ask for additional data, even when the original safety criteria were fully met. The requirements of the Cartagena Protocol and the highly politicized regulatory procedures are largely responsible for the fact that GMOs are almost completely banned in Europe and Africa, with very few exceptions. It is important to stress that many other agricultural technologies that are widely used in conventional and organic farming would not have been approved if they had undergone the same regulatory procedures that are now used for GM crops in many countries.

As I have argued in chapters 3 and 6, the premises on which the Cartagena Protocol and many national regulatory approaches for GMOs build are wrong. In the 1980s and early 1990s, the technology was new and little experience was available about the behavior of foreign transgenes in the plant genome. Many in the wider public did not know how improved crop varieties were developed previously but considered the targeted transfer of genes across species boundary as highly unnatural. This provided a fertile ground for narratives about catastrophic risks. However, the last 30 years of risk research and 20 years of commercial experience have shown that GM crops are not *per se* more dangerous than their conventional counterparts. GM crops have an unblemished safety record. Certain risks of new crop traits are possible, but these are related to the product, not the breeding process. That is, the same risks would also occur if the new traits were developed through conventional breeding. Hence, the approach of singling out GM crops and regulating them differently from other technologies lacks a scientific basis.

Some argue that precaution is still warranted because there may be unknown risks associated with GMOs. But is it really sensible to ban a powerful technology on the basis of potential unknown risks, for which there is no indication? No serious scientist could rule out with 100 percent certainty that safety issues could never occur, but this holds true for every technology, not only GMOs. How should one prove, for instance, that increased long-term consumption of organic food will not lead to health problems? This cannot be proven, but still banning organic food is not seriously considered. There is also no ban on mutagenic varieties, although—compared to GMOs—changes in the genetic makeup of plants are more profound and unpredictable. Mutagenic varieties are widely used in conventional and organic agriculture. Even technologies where there is an indication of possible long-term health risks are not banned when the actual and expected benefits are large. A case in point is the widespread use of smart phones and other mobile electronic devices.

It is also argued sometimes that GM crops are unsustainable, because monocultures are a reality and resistance buildup already occurred in some pest populations. I have explained earlier that these are problems that need to be addressed, but they are not inherent to GMOs. Hence, a ban on GMOs would not be a good idea and could rather aggravate the problems. Monocultures are often observed when growing one crop is much more lucrative for farmers than growing several crops in rotation; this is typically related to wrong policy incentives and lopsided innovation rates. A trend toward monocultures is a typical case of a technology-transcending risk (as opposed to a technology-inherent risk). Technology-transcending risks are best dealt with by altering the external conditions for the better through improved policies and incentive structures. Pest resistance problems also occur with chemical and biological pesticides, as well as with conventionally bred host plant resistance. Resistance problems need to be addressed through improved crop management.

Social risks are also typical cases of technology-transcending risks. If small farms have lesser access to seed markets than large farms because of credit constraints and bad infrastructure conditions, GM crops and other technologies can aggravate income inequality and foster farm size concentration. But rather than banning technology, the better approach to prevent widening disparity would be to address the market failures for smallholders. There are also observed tendencies of concentration in the biotech and seed industries, fostered by powerful technologies, expensive regulations, and IPRs that restrict the freedom-to-operate and are more difficult to handle for smaller companies. Market power is undesirable as it reduces innovation rates and social welfare and makes technologies unnecessarily expensive. But market power is also an issue in other technology-intensive industries, including computer software, the Internet, smart phones, and medical drugs, without anyone seriously suggesting to ban the underlying technologies. Monopolies need to be prevented through anti-trust policies and reducing particular hurdles for small and medium-sized companies.

It is striking that these broader issues that all transformative technologies are associated with lead to public reactions that are so different for GM crops than for other technologies. For most other technologies all sorts of potential problems are either accepted or somehow dealt with through corrective policies. In contrast, for GM crops every single aspect is used as an argument for an outright rejection or for raising regulatory hurdles to a level that is equivalent to a ban. Overregulation has become a real threat for GMO technology and an impediment for sustainable development. Numerous promising GM crop applications got stuck in the thicket of regulation and political uncertainty. Channeling a single new GM technology through the procedure up to commercial approval can take over 10 years and cost 20 million US$ or more in one country alone. Getting the same technology approved in other countries will further add to the cost for the innovator. It is absolutely unsurprising that under these regulatory conditions only multinational companies are able to commercialize GM technologies. And—given the excessive regulatory compliance costs—these multinationals will only focus on large countries, large crops, and traits of large commercial value. It is completely illogical that GM crops, which are not more dangerous than other agricultural technologies, are regulated so differently. This double standard fuels the false public perception that GM crops are a high-risk technology.

The Anti-GMO Protest Industry

In chapter 7, I examined why public perceptions about GM crop risks and benefits differ so widely from the available scientific evidence. I concluded that Western NGOs have mainly been responsible for creating and perpetuating narratives of fear about GMOs. Western NGOs are also spending huge efforts on spreading these narratives to the rest of the world through the classical media, social media, influence on local

NGOs in developing countries, and direct political lobbying. When many of the environmental NGOs started their anti-biotech campaigns in the mid-1990s, they were probably truly concerned about the potential negative environmental and health effects of GMOs. However, when increasing scientific evidence about the safety of GM crops and the benefits for farmers and consumers became available, NGOs were unwilling to change their position and admit that they had been wrong. Instead, they perpetuated the old narratives about environmental and health risks and further added stories about negative social consequences of GM crops, especially for small-scale farmers in developing countries. NGOs are widely considered as groups that are idealistically fighting for the environment, for social justice, and against commercial interests. This is also why their narratives were so successful in establishing the public conviction that GMOs are evil. Individuals and organizations that dare to speak positively about GMOs are regularly stigmatized as stooges of the biotech industry. I have argued that not all NGOs pursue idealistic motives only.

Not to be misunderstood, I am not saying that the world would be a better place without NGOs. NGOs have an important role to play as counterweight to commercial interests and biased policymaking. They also raise public awareness for otherwise neglected issues. But NGOs have gained too much power in some areas. Instead of being one voice in the public debate, NGOs clearly dominate the debate about GMOs. In national and international policymaking processes they are involved as representatives of civil society, but they have no democratic legitimation. For GMOs at least, they do not represent but dictate public opinion. And their business model hardly allows them to correct positions once taken, even when scientific evidence has proven them wrong. It might be useful to consider some kind of ethical code or oversight institution that helps to maintain the important watchdog function of NGOs while reducing trends of obvious public misguidance.

However, NGOs could only be so successful with their anti-GMO propaganda, because many rich country citizens see new technologies in agriculture very critically anyway. Well-fed urbanites in the United States or Europe often do not appreciate the need for global agricultural production increases. Those who can afford consider food produced on local, small-scale and low-tech farms as a fashionable lifestyle. In this postmodern view, productivity-increasing new technologies are seen as a problem rather than a possible solution for sustainable agriculture. It is often not recognized anymore that technological progress in agriculture contributed substantially to economic development in the rich world over the last 150 years. While nowadays high-tech is welcome in most other spheres of life, food should be supernatural and in harmony with ecology. The issue is that the term "natural" is very subjective. As pointed out repeatedly in this book, all forms of agriculture are unnatural, meaning that the types of crops and foods we produce today would not exist without substantial human interference with nature over thousands of years. And who defines that a crop plant with a few added genes taken from nature is less natural than a mutagenic variety, a crop sprayed with chemical pesticides, or the use of copper solutions to control fungal diseases in organic agriculture?

The desire for low-tech ecological farming seems to be more pronounced in Europe than in the United States. This is probably related to differences in population density and culture. In the United States, intensive farming occurs primarily in relatively sparsely populated regions. And when Americans think about nature they often would have pristine national parks in mind. In Europe, however, many live in close proximity to agricultural land. Nature is often associated with diverse agricultural landscapes because pristine wildlife is found much less in Europe. These differences also affect people's attitudes toward land sharing vs. land sparing approaches to agricultural production.

So what is the problem with different preferences? Can Europe not just continue without GMOs, while other regions may decide differently? Well, in a globalized world this is not so simple, because other regions are strongly affected by what Europe does. First, there is a trade link. The EU is a major agricultural importer, so that exporting countries are constrained in the technology choices when they do not want to jeopardize their export markets. Second, through widespread international outreach of European NGOs and media channels, people abroad increasingly adopt the notion that there must be something wrong with GMOs. Third, through international agreements and bilateral assistance programs, European countries supported the establishment of biosafety policies and EU-style regulatory approaches and hurdles. Thus, Europe imposes its opposition to GMOs upon other countries and regions that depend much more on productivity-increasing agricultural technologies. Europe's anti-GMO influence on Africa has been particularly strong (Paarlberg, 2008).

Do We Really Need New Agricultural Technologies?

Some who oppose GMOs argue that new agricultural technologies are unnecessary from a food security perspective. One line of argumentation is that hunger is a distribution problem, so that a focus on productivity-increasing technologies would be misguided. But this view is shortsighted. First, a fairer distribution requires that poor people's incomes are improved. Many of the poor depend on agriculture as their main source of income and employment, and new agricultural technologies—especially those suitable for the small farm sector—were shown to be a powerful tool for poverty reduction. Productivity-increasing technologies are needed not only to increase food availability but also to improve poor people's economic and social access to food. Second, even when globally seen sufficient food is available today, demand increases over time, so that food shortages would soon become reality when sufficient production growth cannot be achieved. Hence, from a dynamic perspective, hunger is certainly both a distribution and a production problem.

A related argumentation is that common food demand projections build on past trends of rising meat consumption and substantial food waste. Further it is argued that meat consumption and food losses should be reduced, which could drastically change the projections. Indeed, if all people would become vegetarians and no food losses and waste occurred, currently produced foods would suffice for more than ten billion people, so that no production increases would be required. But this is an unrealistic scenario. Meat consumption in many developing countries is still quite low. Poor people increase their meat consumption when incomes rise, which is desirable from a nutrition perspective to increase dietary quality and micronutrient intakes. For rich people, less meat consumption would be good, but globally more people are poor than rich. Moreover, behavioral changes take time. More sustainable consumption patterns are definitely required, but this will not reverse the global trend of rising food demand within the next few decades. Unfortunately, in a bipolar world view a demand side focus is often played-off against a supply side focus, and vice versa. In reality, a multiplicity of approaches is required to promote sustainable agricultural development and food security.

But even when the need for production increases is acknowledged, one may ask whether these can only be achieved with new technologies or whether existing technologies may not do as well. Especially in Africa, yield gaps—defined as the difference between actual and exploitable yields with existing technologies—are large. Hence, encouraging a wider use of these existing technologies could also help to increase effective yields substantially. While this is true in general, it should not be used as an argument against also promoting new technologies such as GM crops. GM crops should not be seen as a substitute for existing technologies but a promising addition. In some situations, GM crops may be superior. For instance, in an African context pest-resistant and nutrient use efficient GM crops may allow substantial yield gains without the need to increase chemical pesticide and fertilizer use.

In other sectors, technological leapfrogging has also been quite useful. In large parts of Africa and Asia, poor rural households now benefit substantially from mobile phones, although they never had access to landline phones. It would have been pointless to insist on first serving all rural areas with landline phone infrastructure before allowing mobile technologies to spread.

Can We Trust the Private Sector?

The private sector clearly dominates the development of GM crops. Almost all of the GM crops available so far were commercialized by multinational companies, and many of the interesting GM traits in the pipeline are also in the hands of the private sector. This private sector dominance is an important reason for the widespread public suspicion. It is also a factor that makes it much easier for anti-biotech campaigners to portray GMOs as something that only serves company profits while damaging the environment and hurting social welfare. Especially when it comes to smallholder farmers in developing countries, many feel that private sector seed provision could lead to rising inequality and exploitation. Small farms and large companies, it is felt, do not fit together. In the cognitive shortcut, this leads to rejection of GMOs, as these are likened with Monsanto and other "greedy" multinationals.

But this view is too simplistic. There are many examples where private companies serve poor rural households very well. Private companies of consumer goods, such as soaps, candles, and biscuits, have adjusted their marketing strategies and packaging sizes, so that these products are available in almost every remote village of the world. Mobile phones, private telecommunication services, and related mobile applications offered by private companies are nowadays widely used by smallholder farmers (Kikulwe et al., 2014). In many regions of Africa, over 80 percent of the rural households own a mobile phone. For GM crops, the situation is not much different. As I have shown in chapter 4, millions of smallholders in China, India, and a few other developing countries benefit significantly from GM cotton seeds purchased from Monsanto and other companies.

Private companies can be efficient providers of seeds and other goods and services for poor rural households when markets are competitive (Prahalad, 2004). Market power needs to be avoided, but rejecting new technologies is hardly a good approach to increase competitiveness. As discussed, the opposition to GMOs and the heavy regulation have rather contributed to concentration in the biotech and seed industries. In any case, if the private sector dominance is really felt as a problem in GM crop development, rather than rejecting the technology it would be more logical to call for more public sector research on GM crops as a healthy counterbalance.

Leaving GM crop development entirely in the hands of a few multinationals may indeed be problematic, because these companies focus on markets with large commercial potential. In some cases, the same crop is grown by diverse types of farmers around the world. In these cases, smallholder farmers in developing countries can benefit from spillovers of technologies that were initially developed for larger farms in developed countries. Cases in point are Bt cotton, which is already widely grown by smallholders, and Bt maize, which could be grown more widely by smallholders if it were approved in additional countries of sub-Saharan Africa. Other crops, however, are not widely grown by large commercial farms, although they are of major importance for smallholders. Examples are sorghum, millet, teff, cassava, sweetpotato, and various pulses. Research on such orphan crops needs to be carried out by the public sector. This is also important to prevent further erosion of crop species diversity through GM technology. Hence, public sector GM crop research needs to be increased.

The big advantage of genetic engineering is that the same traits and technologies that multinationals use in commercial crops can also be transferred to orphan crops. Multinationals are often willing to donate some of their technologies for use in humanitarian projects, when market segregation into commercial and non-commercial segments is possible. A few concrete examples of such projects were discussed in chapter 5. The GMO opposition has initially helped to increase the private sector's willingness to donate technologies, as companies hoped that this would contribute to wider GMO acceptance. However, anti-GMO campaigners have clearly overshot with their activities. By deliberately scandalizing cases of adventitious presence of unapproved GM events in seeds and crops, NGOs have tremendously increased reputational risks for multinational companies. As a response, companies have again become more hesitant to donate GM technologies to the public sector, unless experienced international organizations are involved that are capable of handling possible issues of technology stewardship. Technology donations, open-source type of further developments, and public-private partnerships are promising avenues to make the best technologies available to the poor at affordable prices. As the example of Golden Rice has shown (chapter 5), even for crop traits that are not of major commercial interest, companies can and should support GM crop development for humanitarian purposes because of their larger R&D capacities and experience. Such efforts must not be thwarted through unfair anti-biotech campaigns.

The Way Ahead

For millennia, humans have domesticated plants and modified crop plant genomes in order to produce more and better food. The scientific knowledge to improve the genetic makeup of plants has increased, and the methods available to breeders have become much more sophisticated over time. Plant genetic engineering is a powerful set of new methods that have increased the precision of crop improvement and the genetic variability available for developing desirable crop traits. The first GM crops that were developed and commercialized already produced significant benefits. But herbicide-tolerant and insect-resistant crops are only the beginning. Much more interesting GM crop technologies are in the R&D pipeline. These emerging GM crops can help address some of the great challenges of agriculture and natural resource scarcities of the twenty-first century. Not using and further developing these technologies would be irresponsible. I am convinced that sustainable food security will not be feasible without firmly integrating plant genetic engineering into the toolbox of agricultural R&D.

However, many in the wider public are deeply persuaded that GMOs are evil. This misconception builds on limited scientific understanding, false assumptions, and deliberate deception by anti-GMO activists, aided by the mass media and various groups of stakeholders who benefit from this opposition. Considering how successfully GM crop applications were blocked in Europe and many other parts of the world during the last 20 years, one might conclude that there is little hope for improvement. But historically there have been many other breakthrough technologies that were initially met with a lot of public resistance, including the steam train, the automobile, personal computers, and the Internet. Hence, there is still cause for optimism that GM crop technologies will also be accepted eventually. This will require more integrity in the public debate. Anti-GMO activist organizations must stop perpetuating narratives that were shown to be untrue. Every individual has the right to think and say what he/she wants. But NGOs, which are perceived by many as particularly trustworthy, have a moral obligation not to misinform.

The media also has an important role to play in improving the public debate about GMOs. Negative headlines may sell better, but perpetuating narratives and prejudices without further investigation may have to be traded off against the journalist code of ethics. Media reports about scientific topics should be more science-based and less influenced by biased claims of pressure groups.

Politicians should refrain from using the GMO topic for opportunistic statements and policy actions. And scientists working in related fields should engage more seriously in communicating research results to the public. Individual researchers may sometimes be perceived as biased, but science academies and associations often enjoy a high level of reputation. Likewise, international organizations dealing with food and agriculture, such as the FAO and the CGIAR, should communicate about the potentials and limitations of GM crops more proactively. In science outreach activities, journalists, teachers, and school teaching materials should be targeted with priority because of their important multiplier effects.

Like any transformative technology, GMOs raise certain questions that need to be addressed to avoid undesirable side-effects. Some of these questions are rightly raised by biotech critics, but the conclusion that any technology-transcending problem would justify a ban is certainly inappropriate. Unfortunately, the entrenched fundamental debate about banning or allowing GMOs has often overshadowed more detailed questions of suitable technology management. Relevant questions for which institutional adjustments may be required include the following. How can we ensure that GM crops are used sustainably as part of diverse agricultural systems and not as substitutes for proper agronomy? How can market power by a few multinationals be prevented? How can we facilitate the development of GM orphan crops and traits for the benefit of the poor? How can we ensure that suitable GM crop technologies will actually reach the poor through appropriate technology transfer mechanisms? What is the appropriate level of IPR protection in developed and developing countries? Finding answers to these and other relevant questions will require more research and a more constructive public and policy dialogue.

A major stumbling block for the more widespread approval and use of GMOs has been the overregulation of potential biosafety and food safety issues. Overregulation makes GM crop technologies unnecessarily expensive and creates unpredictable delays. Overregulation is closely related to the GMO opposition and the low levels of public acceptance. However, even when public acceptance levels increase, the regulatory barriers will remain, because they build on national and international policies. Hence, profound regulatory reform will be needed. On the one hand, the breeding-process-based regulation observed in Europe and many other countries should be replaced by a product-based regulatory approach. Product-based regulation means that the risks of crop traits are evaluated regardless of which breeding approach was used to develop these traits. On the other hand, risk regulation needs to be less politicized. There may well be important arguments other than health and environmental risks. Such other arguments need to be discussed, but not as part of the biosafety and food safety regulatory procedures.

For individual countries, it will be difficult to step out of international agreements like the Cartagena Protocol. But groups of countries that realize that these international agreements are not suitable for their conditions may well encourage followers. As GM crops can be such a powerful technology to reduce poverty and promote food security and sustainable development, developing country researchers and politicians should feel encouraged to break with the European model and play the role of technology champions more proactively. It would also be very useful for the UN to cancel or substantially revise the Cartagena Protocol, adjusting it to more recent scientific evidence.

REFERENCES

Abedullah, Kouser, S., Qaim, M. (2015). Bt cotton, pesticide use and environmental efficiency in Pakistan. *Journal of Agricultural Economics* 66, 66–86.

Academics Review (2015). IARC glyphosate cancer review fails on multiple fronts. http://academicsreview.org (homegape), date accessed July 28, 2015.

Acquaah, G. (2012). *Principles of Plant Genetics and Breeding*. Second Edition. John Wiley & Sons, Chichester.

Aerni, P. (2014). The motivation and impact of organized public resistance against agricultural biotechnology. In: Smyth, S.J., Phillips, P.W.B., Castle, D., eds. *Handbook on Agriculture, Biotechnology and Development*. Edward Elgar, Cheltenham, UK, pp. 256–276.

Alberts, B., Beachy, R., Baulcombe, D., Blobel, G., Datta, S., Fedoroff, N., Kennedy, D., Khush, G.S., Peacock, J., Rees, M., Sharp, P. (2013). Standing up for GMOs. *Science* 341, 1320.

Ali, A., Abdulai, A. (2010). The adoption of genetically modified cotton and poverty reduction in Pakistan. *Journal of Agricultural Economics* 61, 175–192.

Alston, J.M., Kalaitzandonakes, N., Kruse, J. (2014). The size and distribution of the benefits from the adoption of biotech soybean varieties. In: Smyth, S.J., Phillips, P.W.B., Castle, D., eds. *Handbook on Agriculture, Biotechnology and Development*. Edward Elgar, Cheltenham, UK, pp. 728–751.

Alston, J.M., Norton, G.W., Pardey, P.G. (1995). *Science under Scarcity: Principles and Practices of Agricultural Research Evaluation and Priority Setting*. Cornell University Press, Ithaca, NY.

Ammann, K. (2014). Genomic misconception: a fresh look at the biosafety of transgenic and conventional crops. A plea for a process agnostic regulation. *New Biotechnology* 31, 1–17.

Antoniou, M., Brack, P., Carrasco, A., Fagan, J., Habib, M., Kageyama, P., Leifert, C., Nodari, R.O., Pengue, W. (2010). *GM Soy: Sustainable? Responsible?* GLS Bank and ARGE Gentechnik-frei, Vienna.

Apel, A. (2010). The costly benefits of opposing agricultural biotechnology. *New Biotechnology* 27, 635–640.

Arber, W. (2010). Genetic engineering compared to natural genetic variations. *New Biotechnology* 27, 517–521.

Areal, F.J., Riesgo, L., Rodríguez-Cerezo, E. (2013). Economic and agronomic impact of commercialized GM crops: a meta-analysis. *Journal of Agricultural Science* 151, 7–33.

Arjó, G., Portero, M., Pinol, C., Vinas, J., Matias-Guiu, X., Capell, T., Bartholomaeus, A., Parrott, W., Christou, P. (2013). Plurality of opinion, scientific discourse and pseudoscience: an in depth analysis of the Séralini et al. study claiming that Roundup Ready corn or the herbicide Roundup cause cancer in rats. *Transgenic Research* 22, 255–267.

Bagla, P. (2012). Negative report on GM crops shakes government's food agenda. *Science* 337, 789.

Barker, G. (2006). *The Agricultural Revolution in Prehistory: Why Did Foragers Become Farmers?* Oxford University Press, New York.

Barrows, G., Sexton, S., Zilberman, D. (2014). Agricultural biotechnology: the promise and prospects of genetically modified crops. *Journal of Economic Perspectives* 28, 99–120.

Bartsch, D., Devos, Y., Hails, R., Kiss, J., Krogh, P.H., Mestdagh, S., Nuti, M., Sessitsch, A., Sweet, J., Gathmann, A. (2010). Environmental impact of genetically modified maize expressing cry1 proteins. In: Kempken, F., Jung, C., eds. *Genetic Modification of Plants: Agriculture, Horticulture and Forestry.* Springer, Berlin, pp. 575–614.

Basu, A.K., Qaim, M. (2007). On the adoption of genetically modified seeds in developing countries and the optimal types of government intervention. *American Journal of Agricultural Economics* 89, 784–804.

Baulcombe, D., Dunwell, J., Jones, J., Pickett, J., Puigdomenech, P. (2013). GM Science Update. A Report to the Council for Science and Technology, London.

Becker, H. (2011). *Pflanzenzüchtung.* Second Edition, Ulmer, Stuttgart.

Beckmann, V., Soregaroli, C., Wesseler, J. (2006). Coexistence rules and regulations in the European Union. *American Journal of Agricultural Economics* 88, 1193–1199.

Bennett, R., Kambhampati, U., Morse, S., Ismael, Y. (2006). Farm-level economic performance of genetically modified cotton in Maharashtra, India. *Review of Agricultural Economics* 28, 59–71.

Bennett, R., Morse, S., Ismael, Y. (2003). Bt cotton, pesticides, labour and health: a case study of smallholder farmers in the Makhathini Flats, Republic of South Africa. *Outlook on Agriculture* 32, 123–128.

BfR (2015). EU-Wirkstoffprüfung zu Glyphosat: Stand der Dinge und Ausblick. BfR Mitteilung 002/2015, Bundesinstitut für Riskobewertung (German Federal Institute for Risk Assessment), http://www.bfr.bund.de (homepage), date accessed July 28, 2015.

Blumler, M., Byrne, R. (1991). The ecological genetics of domestication and the origins of agriculture. *Current Anthropology* 32, 23–54.

Borlaug, N. (2007). Feeding a hungry world. *Science* 318, 359.

Borlaug, N. (2000). The Green Revolution Revisited and the Road Ahead. Special 30th Anniversary Lecture, Norwegian Nobel Institute, Oslo.

Brookes, G., Barfoot, P. (2014). Key global environmental impacts of genetically modified (GM) crop use 1996–2012. *GM Crops & Food* 5(2), 149–160.

Brookes, G., Barfoot, P. (2013). The global income and production effects of genetically modified (GM) crops 1996–2011. *GM Crops & Food* 4(1), 74–83.

Brown, L. (2012). *Full Planet, Empty Plates: The New Geopolitics of Food Scarcity.* W.W. Norton & Company, New York.

Brush, S.B., ed. (2000). *Genes in the Field: On-Farm Conservation of Crop Diversity.* Lewis Publishers, New York.

Busch, R.J. (2010). Public perceptions of modern biotechnology and the necessity to improve communication. In: Kempken, F., Jung, C., eds. *Genetic Modification of Plants: Agriculture, Horticulture and Forestry.* Springer, Berlin, pp. 649–661.

Carpenter, J.E. (2010). Peer-reviewed surveys indicate positive impact of commercialized GM crops. *Nature Biotechnology* 28, 319–321.

Carter, C.A., Moschini, G., Sheldon, I., eds. (2011). *Genetically Modified Food and Global Welfare.* Emerald Group Publishing, Bingley, UK.

Casassus, B. (2013). Study linking GM maize to rat tumours is retracted. *Nature,* doi: 10.1038/nature.2013.14268.

CIMMYT (2015). Water Efficient Maize for Africa. International Maize and Wheat Improvement Center, http://www.cimmyt.org (homepage), date accessed April 20, 2015.

Coalition for a GM-Free India (2012). 10 Years of Bt Cotton: False Hypes and False Promises: Cotton Farmers' Crisis Continues with Crop Failures and Suicides. Coalition for a GM-Free India, http://indiagminfo.org (homepage), date accessed April 20, 2015.

Cohen, W., Levin, R. (1989). Empirical studies of innovation and market structure. In: Schmalensee, R., Willig, R., eds. *Handbook of Industrial Organization,* Vol. 2. Elsevier, Amsterdam, pp. 1059–1107.

Cordell, D., Drangert, J.O., While S. (2009). The story of phosporous: global food security and food for thought. *Global Environmental Change* 19, 292–305.

Cordell, D., White, S. (2015). Tracking phosphorus security: indicators of phosphorus vulnerability in the global food system. *Food Security* 7, 337–350.

Cossani, C.M., Reynolds, M.P. (2012). Physiological traits for improving heat tolerance in wheat. *Plant Physiology* 160, 1710–1718.

Cressey, D. (2013). A new breed. *Nature* 497, 27–29.

Crisp, A., Boschetti, C., Perry, M., Tunnacliffe, A., Micklem, G. (2015). Expression of multiple horizontally acquired genes is a hallmark of both vertebrate and invertebrate genomes. *Genome Biology* 15, 50.

Crost, B., Shankar, B., Bennett, R., Morse, S. (2007). Bias from farmer self selection in genetically modified crop productivity estimates: evidence from Indian data. *Journal of Agricultural Economics* 58, 24–36.

Curtis, K.R., McCluskey, J.J., Swinnen, J.F.M. (2008). Differences in global risk perceptions of biotechnology and the political economy of the media. *International Journal of Global Environmental Issues* 8, 77–89.

Darwin, C. (1876). *The Origin of Species by Means of Natural Selection, or the Preservation of Favoured Races in the Struggle for Life.* Sixth Edition, John Murray, London.

Davidson, S.N. (2008). Forbidden fruit: transgenic papaya in Thailand. *Plant Physiology* 147, 487–493.

Davison, J. (2010). GM plants: science, politics and EC regulations. *Plant Science* 178, 94–98.

De Steur, H., Blancquaert, D., Strobbe, S., Lambert, W., Gellynck, X., Van Der Straeten, D. (2015). Status and market potential of transgenic biofortified crops. *Nature Biotechnology* 33, 25–29.

De Steur, H., Gellynck, X., Van Der Straeten, D., Lambert, W., Blancquaert, D., Qaim, M. (2012). Potential impact and cost-effectiveness of multi-biofortified rice in China. *New Biotechnology* 29, 432–442.

DeFrancesco, L. (2013). How safe does transgenic food need to be? *Nature Biotechnology* 31, 794–802.

Devos, Y., Dillen, K., Demont, M. (2014). How can flexibility be integrated into coexistence regulations? A review. *Journal of the Science of Food and Agriculture* 94, 381–387.

Devos, Y., Lheureux, K., Schiemann, J. (2010). Regulatory oversight and safety assessment of plants with novel traits. In: Kempken, F., Jung, C., eds. *Genetic Modification of Plants: Agriculture, Horticulture and Forestry*. Springer, Berlin, pp. 553–574.

DFG (2010). *Grüne Gentechnik*. Deutsche Forschungsgemeinschaft, Wiley-VCH, Weinheim.

Diamond, J. (1999). *Guns, Germs and Steel: The Fates of Human Society*. W.W. Norton & Company, New York.

Dubock, A. (2014). The politics of Golden Rice. *GM Crops & Food* 5(3), 210–222.

Duvick, D. (1986). Plant breeding: past achievements and expectations for the future. *Economic Botany* 40, 289–297.

EASAC (2013). Planting the Future: Opportunities and Challenges for Using Crop Genetic Improvement Technologies for Sustainable Agriculture. EASAC Policy Report 21, European Academies Science Advisory Council, Halle.

Ecker, O., Qaim, M. (2011). Analyzing nutritional impacts of policies: an empirical study for Malawi. *World Development* 39, 412–428.

Eicher, C.K., Staatz, J., eds. (1998). *International Agricultural Development*. Third Edition, Johns Hopkins University Press, Baltimore, MD.

Eisenstein, M. (2014). Against the grain. *Nature* 514, S55–S57.

ENSSER (2015). European Network of Scientists for Social and Environmental Responsibility, http://www.ensser.org (homepage), date accessed April 20, 2015.

ENSA (2015). Engineering Nitrogen Symbiosis for Africa, https://www.ensa.ac.uk.

European Commission (2010a). *Europeans and Biotechnology in 2010*. Directorate-General for Research and Innovation, European Commission, Brussels.

European Commission (2010b). *A Decade of EU-Funded GMO Research, 2001–2010*. Directorate-General for Research and Innovation, European Commission, Brussels.

European Parliament (2015). Parliament backs GMO opt-out for EU member states. Press Release, January 13, 2015, Brussels.

Evenson, R.E., Gollin D. (2003). Assessing the impact of the green revolution, 1960–2000. *Science* 300, 758–762.

Evenson, R.E., Raney, T., eds. (2007). *The Political Economy of Genetically Modified Foods*. Edward Elgar, Cheltenham, UK.

Fagerström, T., Dixelius, C., Magnusson, U., Sundström, J.F. (2012). Stop worrying; start growing. *EMBO Reports* 13, 493–497.

Falck-Zepeda, J., Gruère, G., Sithole-Niang, I., eds. (2013). *Genetically Modified Crops in Africa: Economic and Policy Lessons from Countries South of the Sahara*. International Food Policy Research Institute, Washington, DC.

Falkner, R., ed. (2007). *The International Politics of Genetically Modified Food*. Palgrave Macmillan, New York.

Fan, S., Chan-Kang, C., Qian, K., Krishnaiah, K. (2005). National and international agricultural research and rural poverty: the case of rice research in India and China. *Agricultural Economics* 33, 369–379.

FAO (2015a). *The State of Food Insecurity in the World 2015*. Food and Agriculture Organization of the United Nations, Rome.

FAO (2015b). FAOSTAT Production Data. Food and Agriculture Organization of the United Nations, http://faostat.fao.org/site/291/default.aspx, date accessed April 20, 2015.

FAO (2013). *Climate-Smart Agriculture Sourcebook*. Food and Agriculture Organization of the United Nations, Rome.

FAO (2011). *The State of Food Insecurity in the World: How Does International Price Volatility affect Domestic Economies and Food Security?* Food and Agriculture Organization of the United Nations, Rome.

FAO (2009). *Responding to the Challenges of a Changing World: The Role of New Plant Varieties and High Quality Seed in Agriculture*. Proceedings of the Second World Seed Conference, Food and Agriculture Organization of the United Nations, Rome.

Fernandez-Cornejo, J. (2004). *The Seed Industry in US Agriculture: An Exploration of Data and Information on Crop Seed Markets, Regulation, Industry Structure, and Research and Development*. Agricultural Information Bulletin 786, Economic Research Service, US Department of Agriculture, Washington, DC.

Fernandez-Cornejo, J., Hendricks, C., Mishra, A. (2005). Technology adoption and off-farm household income: the case of herbicide-tolerant soybeans. *Journal of Agricultural and Applied Economics* 37, 549–563.

Fernandez-Cornejo, J., Wechsler, J.J., Livingston, M., Mitchell, L. (2014). Genetically Engineered Crops in the United States. Economic Research Report ERR-162, United Sates Department of Agriculture, Washington, DC.

FiBL, IFOAM (2014). *The World of Organic Agriculture 2014*. Research Institute of Organic Agriculture, International Federation of Organic Agriculture Movements, Frick and Bonn.

Finger, R., El Benni, N., Kaphengst, T., Evans, C., Herbert, S., Lehmann, B., Morse, S., Stupak, N. (2011). A meta-analysis on farm-level costs and benefits of GM crops. *Sustainability* 3, 743–762.

Folcher, L., Delos, M., Marengue, E., Jarry, M., Weissenberger, A., Eychenne, N., Regnault-Roger, C. (2010). Lower mycotoxin levels in Bt maize grain. *Agronomy for Sustainable Development* 30, 711–719.

Foresight (2011). *The Future of Food and Farming: Challenges and Choices for Global Sustainability*. The Government Office for Science, London.

Frisvold, G.B., Reeves, J.M. (2014). Aggregate effects: adopters and non-adopters, investors and consumers. In: Smyth, S.J., Phillips, P.W.B., Castle, D., eds. *Handbook on Agriculture, Biotechnology and Development*. Edward Elgar, Cheltenham, UK, pp. 691–715.

Gaj, T., Gersbach, C.A., Barbas, C.F. (2013). ZFN, TALEN, and CRISPR/Cas-based methods for genome engineering. *Trends in Biotechnology* 31, 397–405.

General Court of the European Union (2013). The General Court has annulled the Commission's decisions concerning authorisation to place on the market the genetically modified potato Amflora. Press release, December 13, 2013, Luxembourg.

Giddings, L.V., Matthew, S., Caine, M. (2013). *Feeding the Planet in a Warming World: Building Resilient Agriculture through Innovation*. London School of Economics and Political Science, London.

Gilbert, N. (2013). A hard look at GM crops. *Nature* 497, 24–26.

Glover D. (2010). Is Bt cotton a pro-poor technology? A review and critique of the empirical record. *Journal of Agrarian Change* 10, 482–509.

GMO Compass (2015). GMO Database, http://www.gmo-compass.org/eng/gmo/db, date accessed April 20, 2015.

Godfray, H.C.J. (2015). The debate over sustainable intensification. *Food Security* 7, 199–208.

Godfray, H.C.J., Beddington, J.R., Crite, I.R., Haddad, L., Lawrence, D., Muir, J.F., Pretty, J., Robinson, S., Thomas, S.M., Toulmin, C. (2010). Food security: the challenge of feeding 9 billion people. *Science*, 327, 812–818.

Gonsalves, C.V., Gonsalves, D. (2014). The Hawaii papaya story. In: Smyth, S.J., Phillips, P.W.B., Castle, D., eds. *Handbook on Agriculture, Biotechnology and Development*. Edward Elgar, Cheltenham, UK, pp. 642–660.

González, C., Johnson, N., Qaim, M. (2009). Consumer acceptance of second-generation GM foods: the case of biofortified cassava in the north-east of Brazil. *Journal of Agricultural Economics* 60, 604–624.

Graff, G.D., Hochman, G., Zilberman D. (2009). The political economy of agricultural biotechnology policies. *AgBioForum* 12, 34–46.

Graff, G.D., Phillips, D., Lei, Z., Oh, S., Nottenburg, C., Pardey, P.G. (2013). Not quite a myriad of gene patents. *Nature Biotechnology* 31, 404–410.

Graham, S. (2002). Journal retracts support for claims of invasive GM corn. *Scientific American*, http://www.scientificamerican.com/article/journal-retracts-support, date accessed April 20, 2015.

Greenpeace (2015). How Greenpeace got China to say no to GE rice, http://www.greenpeace.org/eastasia/specials/gpm01/greenpeace-china-no-gerice, date accessed April 20, 2015.

Greenpeace (2013). *Golden Illusion: The Broken Promises of "Golden" Rice.* Greenpeace International, Amsterdam.

Gruère, G., Sengupta, D. (2011). Bt cotton and farmer suicides in India: an evidence-based assessment. *Journal of Development Studies* 47, 316–337.

Hainzelin, E. (2013). *Cultivating Biodiversity to Transform Agriculture.* Springer, New York.

Hareau, G., Norton, G.E., Mills, B.F., Peterson, E. (2005). Potential benefits of transgenic rice in Asia: a general equilibrium analysis. *Quarterly Journal of International Agriculture* 44, 229–246.

Harlan, J. (1971). Agricultural origins: centers and noncenters. *Science* 174, 468–474.

Harmon, A. (2013). A race to save the orange by altering its DNA. *New York Times*, July 28, 2013, p. A1.

Hawkesford, M.J. (2012). Improving nutrient use in crops. *eLS*, doi: 10.1002/9780470015902.a0023734.

Hazell, P., Ramasamy, C. (1991). *The Green Revolution Reconsidered: The Impact of High-Yielding Rice Varieties in South India.* The Johns Hopkins University Press, Baltimore, MD.

Heisey, P., Schimmelpfennig, D. (2006). Regulation and the structure of biotechnology industries. In: Just, R.E., Alston, J.M., Zilberman, D., eds. *Regulating Agricultural Biotechnology: Economics and Policy.* Springer, New York, pp. 421–436.

Herring, R.J. (2015). State science, risk and agricultural biotechnology: Bt cotton to Bt Brinjal in India. *Journal of Peasant Studies* 42, 159–186.

Herring, R.J. (2010). Epistemic brokerage in the bio-property narrative: contributions to explaining opposition to transgenic technologies in agriculture. *New Biotechnology* 27, 614–622.

Herring, R.J. (2007). Stealth seeds: bioproperty, biosafety, biopolitics. *Journal of Development Studies* 43, 130–157.

Hertel, T. (2015). The challenges of sustainably feeding a growing planet. *Food Security* 7, 185–198.

Hesselbarth, K. (2013). The deadly myth of Bt cotton. *AgriFuture*, 8–10.

Hossain, F., Pray, C., Lu, Y., Huang, J., Fan, C., Hu, R. (2004). Genetically modified cotton and farmers' health in China. *International Journal of Occupational and Environmental Health* 10, 296–303.

Hotz, C., Loechl, C., de Brauw, A., Eozenou, P., Gilligan, D., Moursi, M., Munhaua, B., van Jaarsveld, P., Carriquiry, A., Meenakshi, J.V. (2012). A large-scale intervention to introduce orange sweet potato in rural Mozambique increases vitamin A intakes among children and women. *British Journal of Nutrition* 108, 163–176.

House of Commons (2015). Advanced Genetic Techniques for Crop Improvement: Regulation, Risk and Precaution. Fifth Report of Session 2014–15, Science and Technology Committee, House of Commons, UK Parliament, London.

Huang, J., Hu, R., Rozelle, S., Pray, C. (2005). Insect-resistant GM rice in farmers' fields: assessing productivity and health effects in China. *Science* 308, 688–690.

Huang, J., Hu, R., Qiao, F., Yin, Y., Liu, H., Huang, Z. (2015). Impact of insect-resistant GM rice on pesticide use and farmers' health in China. *Science China Life Sciences*, doi: 10.1007/s11427–014–4768–1.

Huang, J., Pray, C., Rozelle, S. (2002). Enhancing the crops to feed the poor. *Nature* 418, 678–684.

Huffman, W.E. (2011). Contributions of public and private R&D to biotechnology innovation. In: Carter, C.A., Moschini, G., Sheldon, I., eds. *Genetically Modified Food and Global Welfare*. Emerald Group Publishing, Bingley, UK, pp. 115–147.

Huffman, W.E., McCluskey, J.J. (2014). Labeling of genetically modified foods. In: Smyth, S.J., Phillips, P.W.B., Castle, D., eds. *Handbook on Agriculture, Biotechnology and Development*. Edward Elgar, Cheltenham, UK, pp. 467–487.

Hutchison, W.D., Burkness, E.C., Mitchell, P.D., Moon, R.D., Leslie, T.W., Fleischer, S.J., Abrahamson, M., Hamilton, K.L., Steffey, K.L., Gray, M.E., Hellmich, R.L., Kaster, L.V., Hunt, T.E., Wright, R.J., Pecinovsky, K., Rabaey, T.L., Flood, B.R. and Raun, E.S. (2010). Areawide suppression of European corn borer with Bt maize reaps savings to non-Bt maize growers. *Science* 330, 222–225.

IAASTD (2009). *Synthesis Report*. International Assessment of Agricultural Science and Technology for Development, Washington, DC.

IARC (2015). Evaluation of five organophosphate insecticides and herbicides. IARC Monographs 112, International Agency for Research on Cancer, Lyon.

IFPRI (2014). *Global Nutrition Report 2014*. International Food Policy Research Institute, Washington, DC.

IFPRI (2010). *Food Security, Farming, and Climate Change to 2050: Scenarios, Results, Policy Options*, International Food Policy Research Institute, Washington, DC.

IRRI (2015). The C4 Rice Project. International Rice Research Institute, http://irri.org (homepage), date accessed April 20, 2015.

ISAAA (2015). GM Approval Database. International Service for the Acquisition of Agri-Biotech Applications, http://www.isaaa.org/gmapprovaldatabase/default.asp, date accessed April 20, 2015.

ISAAA (2013). Stacked traits in biotech crops. *Pocket K* 42, International Service for the Acquisition of Agri-Biotech Applications, Ithaca, NY.

IWYP (2015). International Wheat Yield Partnership, http://iwyp.org (homepage), date accessed April 20, 2015.

James, C. (2014). Global status of commercialized biotech/GM crops: 2014. *ISAAA Briefs* 49, International Service for the Acquisition of Agri-Biotech Applications, Ithaca, NY.

James, C. (2013). Global status of commercialized biotech/GM crops: 2013. *ISAAA Briefs* 46, International Service for the Acquisition of Agri-Biotech Applications, Ithaca, NY.

Juma, C. (2011). *The New Harvest: Agricultural Innovation in Africa.* Oxford University Press, New York.

Just, R.E., Alston, J.M., Zilberman, D., eds. (2006). *Regulating Agricultural Biotechnology: Economics and Policy.* Springer, New York.

Kahneman, D. (2011). *Thinking, Fast and Slow.* Macmillan, New York.

Kalaitzandonakes, N., Alston, J.M., Bradford, K.J. (2007). Compliance costs for regulatory approval of new biotech crops. *Nature Biotechnology* 25, 509–511.

Kalaitzandonakes, N., Kaufman, J., Miller, D. (2014). Potential economic impacts of zero thresholds for unapproved GMOs: the EU case. *Food Policy* 45, 146–157.

Kathage, J., Qaim, M. (2012). Economic impacts and impact dynamics of Bt (*Bacillus thuringiensis*) cotton in India. *Proceedings of the National Academy of Sciences USA* 109, 11652–11656.

Kempken, F., Jung, C., eds. (2010). *Genetic Modification of Plants: Agriculture, Horticulture and Forestry.* Springer, Berlin.

Kerr, W.A. (2014). The trade system and biotechnology. In: Smyth, S.J., Phillips, P.W.B., Castle, D., eds. *Handbook on Agriculture, Biotechnology and Development.* Edward Elgar, Cheltenham, UK, pp. 217–229.

Kikulwe, E.M., Fischer, E., Qaim, M. (2014). Mobile money, smallholder farmers, and household welfare in Kenya. *PLOS ONE* 9, e109804.

Kloppenburg, J. (2004). *First the Seed: The Political Economy of Plant Biotechnology.* Second Edition. University of Wisconsin Press, Madison.

Klümper, W., Qaim, M. (2014). A meta-analysis of the impacts of genetically modified crops. *PLOS ONE* 9, e111629.

Kostandini, G., Mills, B.F., Omamo, S.W., Wood, S. (2009). Ex ante analysis of the benefits of transgenic drought tolerance research on cereal crops in low-income countries. *Agricultural Economics* 40, 477–492.

Kouser, S., Qaim, M. (2014). Bt cotton, damage control, and optimal levels of pesticide use in Pakistan. *Environment and Development Economics* 19, 704–723.

Kouser, S., Qaim, M. (2013). Valuing financial, health, and environmental benefits of Bt cotton in Pakistan. *Agricultural Economics* 44, 323–335.

Kouser, S., Qaim, M. (2011). Impact of Bt cotton on pesticide poisoning in smallholder agriculture: a panel data analysis. *Ecological Economics* 70, 2105–2113.

Krishna, V.V., Qaim, M. (2012). Bt cotton and sustainability of pesticide reductions in India. *Agricultural Systems* 107, 47–55.

Krishna, V.V., Qaim, M. (2008a). Potential impacts of Bt eggplant on economic surplus and farmers' health in India. *Agricultural Economics* 38, 167–180.

Krishna, V.V., Qaim, M. (2008b). Consumer attitudes toward GM food and pesticide residues in India. *Review of Agricultural Economics* 30, 233–251.

Krishna, V., Qaim, M., Zilberman, D. (2015). Transgenic crops, production risk, and agrobiodiversity. *European Review of Agricultural Economics*, doi: 10.1093/erae/jbv012.

Krishnan, M. (2015). Harvesting controversy with GM crops in India, http://www.dw.de/harvesting-controversy-with-gm-crops-in-india/a-18314484.

Kuntz, M. (2014). "Parallel science" of NGO advocacy groups: how post-modernism encourages pseudo-science. Genetic Literacy Project, http://geneticliteracyproject.org (homepage), date accessed April 20, 2015.

Kuzma, J., Kokotovich, A. (2011). Renegotiating GM crop regulation: targeted gene-modification technology raises new issues for the oversight of genetically modified crops. *EMBO Reports* 12, 883–888.

Kyndt, T., Quispe, D., Zhai, H., Jarret, R., Ghislain, M., Liu, Q., Gheysen, G., Kreuze, J.F. (2015). The genome of cultivated sweet potato contains *Agrobacterium* T-DNAs with expressed genes: an example of a naturally transgenic food crop. *Proceedings of the National Academy of Sciences USA*, doi: 10.1073/pnas.1419685112.

La Via Campesina (2015). International Peasant's Movement, http://viacampesina.org (homepage), date accessed April 20, 2015.

Laursen, L. (2013). Greenpeace campaign prompts Philippine ban on Bt eggplant trials. *Nature Biotechnology* 31, 777–778.

Lee, D. (2005). Agricultural sustainability and technology adoption: issues and policies for developing countries. *American Journal of Agricultural Economics* 87, 1325–1334.

Leisinger, K.M. (1995). Sociopolitical effects of new biotechnologies in developing countries. *Food Agriculture, and the Environment Discussion Paper* 2, Vision 2020, International Food Policy Research Institute, Washington, DC.

Lipton, M., Longhurst, R. (1989). *New Seeds and Poor People*. Johns Hopkins University Press, Baltimore, MD.

Loder, N. (1999). Royal Society: GM food hazard claim is "flawed." *Nature* 399, 188.

Lomborg, B., ed. (2012). *How to Spend $75 Billion to Make the World a Better Place*. Copenhagen Consensus Center, Copenhagen.

Long, S.P., Marshall-Colon, A., Zhu, X.G. (2015). Meeting the global food demand of the future by engineering crop photosynthesis and yield potential. *Cell* 161, 56–66.

Lu, Y., Wu, K., Jiang, Y., Xia, B., Li, P., Feng, H., Wyckhuys, K.A.G., Guo, Y. (2010). Mirid bug outbreaks in multiple crops correlated with wide-scale adoption of Bt cotton in China. *Science* 328, 1151–1154.

Lusk, J.L. (2003). Effect of cheap talk on consumer willingness to pay for Golden Rice. *American Journal of Agricultural Economics* 85, 840–856.

Lynch, D., Vogel, D. (2001). The regulation of GMOs in Europe and the United States: a case-study of contemporary European regulatory politics. Council on Foreign Relations, http://www.cfr.org (homepage), date accessed April 23, 2015.

Lynas, M. (2013). Lecture to Oxford Farming Conference, 3 January 2013, http://www.marklynas.org (homepage), date accessed April 20, 2015.

Mangrauthia, S.K., Revathi, P., Agarwal, P., Sing, A.K., Bhadana, V.P. (2014). Breeding and transgenic approaches for development of abiotic stress tolerance in rice. In: Ahmad, P., Wani, M.R., Azooz, M.M., Tran, L.S.P., eds. *Improvement of Crops in the Era of Climatic Changes*. Springer, New York, pp. 153–190.

Mannion, A. (1995). *The Origins of Agriculture: An Appraisal.* University of Reading, Department of Geography, George Overs, London.

March against Monsanto (2015). http://www.march-against-monsanto.com (homepage), date accessed April 20, 2015.

Marvier, M., McCreedy, C., Regetz, J., Kareiva, P. (2007). A meta-analysis of effects of Bt cotton and maize on nontarget invertebrates. *Science* 316, 1475–1477.

Mayr, E. (1982). *The Growth of Biological Thought: Diversity, Evolution, and Inheritance.* Harvard University Press, Cambridge.

Moore, P. (2010). *Confessions of a Greenpeace Dropout: The Making of a Sensible Environmentalist.* Beatty Street Publishing, Vancouver.

Moschini, G. (2008). Biotechnology and the development of food markets: retrospect and prospects. *European Review of Agricultural Economics* 35, 331–355.

Moschini, G., Lapan, H. (1997). Intellectual property rights and the welfare effects of agricultural R&D. *American Journal of Agricultural Economics* 79, 1229–1242.

Murphy, D. (2007a). *People, Plants and Genes: The Story of Crops and Humanity.* Oxford University Press, New York.

Murphy, D. (2007b). *Plant Breeding and Biotechnology: Societal Context and the Future of Agriculture.* Cambridge University Press, New York.

Navdanya (2004). *Monsanto, the Gene Giant: Peddling Life Sciences or Death Sciences?.* RFSTE, Navdanya, Polaris Institute, New Delhi.

Nisbet, M. (2013). Greenpeace Inc. The Breakthrough Institute, http://the-breakthrough.org/index.php/voices/the-public-square/greenpeace-inc, date accessed April 20, 2015.

Noltze, M., Schwarze, S., Qaim, M. (2013). Impacts of natural resource management technologies on agricultural yield and household income: the system of rice intensification in Timor Leste. *Ecological Economics* 85, 59–68.

Oerke, E.-C. (2006). Crop losses to pests. *Journal of Agricultural Science* 144, 31–43.

Okeno, J.A., Wolt, J.D., Misra, M.K., Rodriguez, L. (2013). Africa's inevitable walk to genetically modified (GM) crops: opportunities and challenges for commercialization. *New Biotechnology* 30, 124–130.

Oxfam (2011). *Who Will Feed the World: The Production Challenge.* Oxfam Research Report, Boston.

Paarlberg, R. (2014). African non-adopters. In: Smyth, S.J., Phillips, P.W.B., Castle, D., eds. *Handbook on Agriculture, Biotechnology and Development.* Edward Elgar, Cheltenham, UK, pp. 166–175.

Paarlberg, R.L. (2008). *Starved for Science: How Biotechnology Is Being Kept Out of Africa.* Harvard University Press, Cambridge, MA.

Paine, J., Shipton, C.A., Chaggar, S., Howells, R.M., Kennedy, M.J., Vernon, G., Wright, S.Y., Hinchliffe, E., Adams, J.L., Silverstone, A.L., Drake, R. (2005). Improving the nutritional value of Golden Rice through increased pro-vitamin A content. *Nature Biotechnology* 23, 482–487.

Palma, G. (2011). Letter to the editor regarding the article by Paganelli et al. *Chemical Research in Toxicology* 24, 775–776.

Palmgren, M.G., Edenbrandt, A.K., Vedel, S.E., Andersen, M.M., Landes, X., Østerberg, J.T., Falhof, J., Olsen, L.I., Christensen, S.B., Sandøe, P., Gamborg, C., Kappel, K., Thorsen, B.J., Pagh, P. (2015). Are we ready for back-to-nature crop breeding? *Trends in Plant Science* 20, 155–164.

Panjabi, R.K.L. (1997). *The Earth Summit at Rio: Politics, Economics, and the Environment.* Northeastern University Press, Boston.

Parrott, W. (2010). Genetically modified myths and realities. *New Biotechnology* 27, 545–551.

Patel, V., Ramasundarahettige, C., Vijayakumar, L., Thakur, J.S., Gajalakshmi, V., Gururaj, G., Suraweera, W., Jha, P. (2012). Suicide mortality in India: a nationally representative survey. *The Lancet* 379, 2343–2351.

Pearse, A. (1980). *Seeds of Plenty, Seeds of Want: Social and Economic Implications of the Green Revolution.* Oxford University Press, New York.

Pemsl, D.E., Gutierrez, A.P., Waibel, H. (2008). The economics of biotechnology under ecosystem disruption. *Ecological Economics* 66, 177–183.

Potrykus, I. (2010). Lessons from the "Humanitarian Golden Rice Project": regulation prevents development of public good genetically engineered crops. *New Biotechnology* 27, 467–472.

Potrykus, I. (2001). Golden Rice and beyond. *Plant Physiology* 125, 1157–1161.

Prahalad, C.K. (2004). *The Fortune at the Bottom of the Pyramid: Eradicating Poverty through Profits.* Prentice Hall, Upper Saddle River, NJ.

Qaim, M. (2014). Agricultural biotechnology in India: impacts and controversies. In: Smyth, S.J., Phillips, P.W.B., Castle, D., eds. *Handbook on Agriculture, Biotechnology and Development.* Edward Elgar, Cheltenham, UK, pp. 126–137.

Qaim, M. (2010). Benefits of genetically modified crops for the poor: household income, nutrition, and health. *New Biotechnology* 27, 552–557.

Qaim, M. (2009). The economics of genetically modified crops. *Annual Review of Resource Economics* 1, 665–693.

Qaim, M. (2005). Agricultural biotechnology adoption in developing countries. *American Journal of Agricultural Economics* 87, 1317–1324.

Qaim, M., de Janvry, A. (2005). Bt cotton and pesticide use in Argentina: economic and environmental effects. *Environment and Development Economics* 10, 179–200.

Qaim, M., de Janvry, A. (2003). Genetically modified crops, corporate pricing strategies, and farmers' adoption: the case of Bt cotton in Argentina. *American Journal of Agricultural Economics* 85, 814–828.

Qaim, M., Klümper, W. (2013). Landwirtschaft für die Hungerbekämpfung. *Chemie in unserer Zeit* 47, 318–326.

Qaim, M., Kouser, S. (2013). Genetically modified crops and food security. *PLOS ONE* 8, e64879.

Qaim, M., Krattiger, A.F., von Braun, J., eds. (2000). *Agricultural Biotechnology in Developing Countries: Towards Optimizing the Benefits for the Poor.* Kluwer Academic Publishers, Boston and Dordrecht.

Qaim, M., Stein, A.J., Meenakshi, J.V. (2007). Economics of biofortification. *Agricultural Economics* 37(s1), 119–133.

Qaim, M., Subramanian, A., Naik, G., Zilberman, D. (2006). Adoption of Bt cotton and impact variability: insights from India. *Review of Agricultural Economics* 28, 48–58.

Qaim, M., Subramanian, A., Sadashivappa, P. (2009). Commercialized GM crops and yield. *Nature Biotechnology* 27, 803–804.

Qaim, M., Traxler, G. (2005). Roundup Ready soybeans in Argentina: farm level and aggregate welfare effects. *Agricultural Economics* 32, 73–86.

Qaim, M., Zilberman, D. (2003). Yield effects of genetically modified crops in developing countries. *Science* 299, 900–902.

Qiao, F. (2015). Fifteen years of Bt cotton in China: the economic impact and its dynamics. *World Development* 70, 177–185.

Raney, T., Matuschke, I. (2011). Current and potential farm-level impacts of genetically modified crops in developing countries. In: Carter, C.A., Moschini, G., Sheldon, I., eds. *Genetically Modified Food and Global Welfare.* Emerald Group Publishing, Bingley, UK, pp. 55–82.

Rao, N.C., Dev, M. (2010). *Biotechnology in Indian Agriculture: Potential, Performance and Concerns.* Center for Economic and Social Studies, Hyderabad.

Raven, P.H. (2010). Does the use of transgenic plants diminish or promote biodiversity? *New Biotechnology* 27, 528–533.

Reuters (2013). Field trials of new nitrogen use efficient rice show increased productivity, leading to increased food security and reduced fertilizer dependence. Press release, 10 September 2013, http://www.reuters.com (homepage), date accessed April 20, 2015.

Reynolds, M., ed. (2010). *Climate Change and Crop Production.* CABI, Wallingford, UK.

Ricroch, A., Hénard-Damave, M.C. (2015). Next biotech plants: new traits, crops, developers and technologies for addressing global challenges. *Critical Reviews in Biotechnology*, doi: 10.3109/07388551.2015.1004521.

Romeis, J., Meissle, M., Alvarez-Alfageme, F., Bigler, F., Bohan, D.A., Devos, Y., Malone, L.A., Pons, X., Rauschen, S. (2014). Potential use of an arthropod database to support the non-target risk assessment and monitoring of transgenic plants. *Transgenic Research* 23, 995–1013.

Romeis, J., Shelton, A.S., Kennedy, G.G., eds. (2008). *Integration of Insect-Resistant Genetically Modified Crops within IPM Programs.* Springer, Berlin.

Ronald, P.C., Adamchak, R.W. (2008). *Tomorrow's Table: Organic Farming, Genetics, and the Future of Food.* Oxford University Press, New York.

Rosegrant, M.W., Koo, J., Cenacchi, N., Ringler, C., Roberston, R., Fisher, M., Cox, C., Garrett, K., Perez, N.D., Sabbagh, P. (2014). *Food Security in a World of Natural Resource Scarcity: The Role of Agricultural Technologies.* International Food Policy Research Institute, Washington, DC.

Rosen, J.B., Schulkin, J. (1998). From normal fear to pathological anxiety. *Psychological Review* 105, 325–350.

Royal Society (2009). *Reaping the Benefits: Science and the Sustainable Intensification of Global Agriculture.* The Royal Society, London.

Ryan, C.D., McHughen, A. (2014). Tomatoes, potatoes and flax: exploring the cost of lost innovation. In: Smyth, S.J., Phillips, P.W.B., Castle, D., eds. *Handbook on Agriculture, Biotechnology and Development*. Edward Elgar, Cheltenham, UK, pp. 841–852.

Schroeder, J.I., Delhaize, E., Frommer, W.B., Guerinot, M.L., Harrison, M.J., Herrera-Estrella, L., Horie, T., Kochian, L.V., Munns, R., Nishizawa, N.K., Tsay, Y.F., Sanders, D. (2013). Using membrane transporters to improve crops for sustainable food production. *Nature* 497, 60–66.

Schubert, D., Williams, D. (2006). "Cisgenic" as a product designation. *Nature Biotechnology* 24, 1327–1329.

Sears, M.K., Hellmich, R.L., Stanley-Horn, D.E., Oberhauser, K.S., Pleasants, J.M., Mattila, H.R., Siegfried, B.D., Dively, G.P. (2001). Impact of Bt corn pollen on monarch butterfly populations: a risk assessment. *Proceedings of the National Academy of Sciences USA* 98, 11937–11942.

Service, R.-F. (2007). A growing threat down on the farm. *Science* 316, 1114–1117.

Sexton, S., Zilberman, D. (2012). Land for food and fuel production: the role of agricultural biotechnology. In: Zivin, J.S.G., Perloff, J.M., eds. *The Intended and Unintended Effects of US Agricultural and Biotechnology Policies*. University of Chicago Press, Chicago, pp. 269–288.

Shiva, V., Barker, D., Lockhart, C. (2011). *The GMO Emperor Has No Clothes*. Navdanya International, New Delhi.

Siegrist, M., Earle, T., Gutscher, H., eds. (2007). *Trust in Cooperative Risk Management: Uncertainty and Skepticism in the Public Mind*. Earthscan, London.

Smale, M., Zambrano, P., Gruère, G., Falck-Zepeda, J., Matuschke, I., Horna, D., Nagarajan, L., Yerramareddy, I., Jones, H. (2009). *Measuring the Economic Impacts of Transgenic Crops in Developing Agriculture during the First Decade: Approaches, Findings, and Future Directions*. International Food Policy Research Institute, Washington, DC.

Smyth, S.J., Phillips, P.W.B., Castle, D., eds. (2014). *Handbook on Agriculture, Biotechnology and Development*. Edward Elgar, Cheltenham, UK.

Snell, C., Bernheim, A., Bergé, J.-P., Kuntz, M., Pascal, G., Paris, A., Ricroch, A.E. (2012). Assessment of the health impact of GM plant diets in long-term and multigenerational animal feeding trials: a literature review. *Food and Chemical Toxicology* 50, 1134–1148.

Stein, A.J., Sachdev, H.P.S., Qaim, M. (2008). Genetic engineering for the poor: Golden Rice and public health in India. *World Development* 36, 144–158.

Stein, A.J., Qaim, M. (2007). The human and economic cost of hidden hunger. *Food and Nutrition Bulletin* 28, 125–134.

Stokstad, E. (2008). Dueling visions for a hungry world. *Science* 319, 1474–1476.

Stone, G.D. (2012). Constructing facts: Bt cotton narratives in India. *Economic & Political Weekly* XLVII (38), 62–70.

Stone, G.D. (2011). Field versus farm in Warangal: Bt cotton, higher yields, and larger questions. *World Development* 39, 387–398.

Subramanian, A., Qaim, M. (2010). The impact of Bt cotton on poor households in rural India. *Journal of Development Studies* 46, 295–311.

Tabashnik, B.E., Brévault, T., Carrière, Y. (2013). Insect resistance to Bt crops: lessons from the first billion acres. *Nature Biotechnology* 31, 510–521.

Tabashnik, B.E., Gassmann, A.J., Crowder, D.W., Carrière, Y. (2008). Insect resistance to Bt crops: evidence versus theory. *Nature Biotechnology* 26, 199–202.

Tan, S.Y., Evans, R.R., Dahmer, M.L., Singh, B.K., Shaner, D.L. (2005). Imidazolinone-tolerant crops: history, current status and future. *Pest Management Science* 61, 246–257.

Tang, G., Hu, Y., Yin, S., Wang, Y., Dallal, G.E., Grusak, M.A., Russell, R.M. (2012). Beta-carotene in Golden Rice is as good as beta-carotene in oil at providing vitamin A to children. *American Journal of Clinical Nutrition* 96, 658–664.

Tang, G., Qin, J., Dolnikowski, G.G., Russell, R.M., Grusak, M.A. (2009). Golden Rice is an effective source of vitamin A. *American Journal of Clinical Nutrition* 89, 1776–1783.

Taverne, D. (2007). The real GM food scandal. *Prospect*, November 2007, 24–27.

Tester, M., Langridge, P. (2010). Breeding technologies to increase crop production in a changing world. *Science* 327, 818–822.

The Economist (2014). Genetically modified crops: field research. *The Economist*, November 8, 2014, 74.

The Economist (2013). Genetically modified crops: food fight. *The Economist*, December 14, 2013, 53.54.

Thirtle, C., Lin, L., Piesse, J. (2003). The impact of research-led agricultural productivity growth on poverty reduction in Africa, Asia and Latin America. *World Development* 31, 1959–1975.

Thomson, J. (2014). Africa. In: Smyth, S.J., Phillips, P.W.B., Castle, D., eds. *Handbook on Agriculture, Biotechnology and Development*. Edward Elgar, Cheltenham, UK, pp. 99–111.

Townsend, J.A., Wright, D.A., Winfrey, R.J., Fu, F., Maeder, M.L., Joung, J.K., Voytas, D.F. (2009). High-frequency modification of plant genes using engineered zinc-finger nucleases. *Nature* 459, 442–445.

Tripp, R. (1996). Biodiversity and modern crop varieties: sharpening the debate. *Agriculture and Human Values* 13, 48–63.

United Nations (2014). *The Millennium Development Goals Report*. United Nations, New York.

Villoria, N.B., Byerlee, D., Stevenson, J. (2014). The effects of agricultural technological progress on deforestation: what do we really know? *Applied Economic Perspectives and Policy* 36, 211–237.

Vitale, J., Vognan, G., Outtara, M. (2014). Cotton. In: Smyth, S.J., Phillips, P.W.B., Castle, D., eds. *Handbook on Agriculture, Biotechnology and Development*. Edward Elgar, Cheltenham, UK, pp. 604–620.

von Braun, J., Teklu, T., Webb, P. (1998). *Famine in Africa: Causes, Responses, and Prevention*. Johns Hopkins University Press, Baltimore, MD.

Wang, F., Ye, C., Zhu, L., Nie, L., Cui, K., Peng, S., Lin, Y., Huang, J. (2012). Yield differences between Bt transgenic rice lines and their non-Bt counterparts, and its possible mechanisms. *Field Crops Research* 126, 8–15.

Wang, S., Just, D.R., Pinstrup-Anderson, P. (2008). Bt cotton and secondary pests. *International Journal of Biotechnology* 10, 113–121.

Wesseler, J., Zilberman, D. (2014). The economic power of the Golden Rice opposition. *Environment and Development Economics* 19, 724–742.

Wilson, W.W. (2014). Wheat: status, outlook and implications. In: Smyth, S.J., Phillips, P.W.B., Castle, D., eds. *Handbook on Agriculture, Biotechnology and Development.* Edward Elgar, Cheltenham, UK, pp. 752–769.

WHO (2015). Vitamin and Mineral Nutrition Information System. World Health Organization, http://www.who.int/vmnis/database/en, date accessed April 20, 2015.

Wolfenbarger, L.L.R., Naranjo, S.E., Lundgren, J.G., Bitzer, R.J., Watrud, L.S. (2008). Bt crop effects on functional guilds of non-target arthropods: a meta-analysis. *PLOS ONE* 3, e2118.

World Bank (2013). *Implementing Agriculture for Development: World Bank Group Agriculture Action Plan 2013–2015.* World Bank, Washington, DC.

World Bank (2007). *World Development Report 2008: Agriculture for Development.* World Bank, Washington, DC.

Wu, F. (2006). Bt corn's reduction of mycotoxins: regulatory decisions and public opinion. In: Just, R.E., Alston, J.M., Zilberman, D., eds. *Regulating Agricultural Biotechnology: Economics and Policy.* Springer, New York, pp. 179–200.

Wu, K.M., Lu, Y.-H., Feng, H.-Q., Jiang, Y.-Y., Zhao, J.-Z. (2008). Suppression of cotton bollworm in multiple crops in China in areas with Bt toxin-containing cotton. *Science* 321, 1676–1678.

Ye, X., Al-Babili, S., Klöti, A., Zhang, J., Lucca, P., Beyer, P., Potrykus, I. (2000). Engineering the provitamin A (β-carotene) biosynthetic pathway into (carotenoid-free) rice endosperm. *Science* 287, 303–305.

Zilberman, D., Ameden, H., Qaim, M. (2007). The impact of agricultural biotechnology on yields, risks, and biodiversity in low-income countries. *Journal of Development Studies* 43, 63–78.

Zimmermann, R., Qaim, M. (2004). Potential health benefits of Golden Rice: a Philippine case study. *Food Policy* 29, 147–168.

INDEX